Jacques Teller, John R. Lee and Catherine Roussey (Eds.)

Ontologies for Urban Development

Studies in Computational Intelligence, Volume 61

Editor-in-chief
Prof. Janusz Kacprzyk
Systems Research Institute
Polish Academy of Sciences
ul. Newelska 6
01-447 Warsaw
Poland
E-mail: kacprzyk@ibspan.waw.pl

Further volumes of this series can be found on our homepage: springer.com

Vol. 38. Art Lew, Holger Mauch
Dynamic Programming, 2007
ISBN 978-3-540-37013-0

Vol. 39. Gregory Levitin (Ed.)
Computational Intelligence in Reliability Engineering, 2007
ISBN 978-3-540-37367-4

Vol. 40. Gregory Levitin (Ed.)
Computational Intelligence in Reliability Engineering, 2007
ISBN 978-3-540-37371-1

Vol. 41. Mukesh Khare, S.M. Shiva Nagendra (Eds.)
Artificial Neural Networks in Vehicular Pollution Modelling, 2007
ISBN 978-3-540-37417-6

Vol. 42. Bernd J. Krämer, Wolfgang A. Halang (Eds.)
Contributions to Ubiquitous Computing, 2007
ISBN 978-3-540-44909-6

Vol. 43. Fabrice Guillet, Howard J. Hamilton (Eds.)
Quality Measures in Data Mining, 2007
ISBN 978-3-540-44911-9

Vol. 44. Nadia Nedjah, Luiza de Macedo Mourelle, Mario Neto Borges, Nival Nunes de Almeida (Eds.)
Intelligent Educational Machines, 2007
ISBN 978-3-540-44920-1

Vol. 45. Vladimir G. Ivancevic, Tijana T. Ivancevic
Neuro-Fuzzy Associative Machinery for Comprehensive Brain and Cognition Modeling, 2007
ISBN 978-3-540-47463-0

Vol. 46. Valentina Zharkova, Lakhmi C. Jain
Artificial Intelligence in Recognition and Classification of Astrophysical and Medical Images, 2007
ISBN 978-3-540-47511-8

Vol. 47. S. Sumathi, S. Esakkirajan
Fundamentals of Relational Database Management Systems, 2007
ISBN 978-3-540-48397-7

Vol. 48. H. Yoshida (Ed.)
Advanced Computational Intelligence Paradigms in Healthcare, 2007
ISBN 978-3-540-47523-1

Vol. 49. Keshav P. Dahal, Kay Chen Tan, Peter I. Cowling (Eds.)
Evolutionary Scheduling, 2007
ISBN 978-3-540-48582-7

Vol. 50. Nadia Nedjah, Leandro dos Santos Coelho, Luiza de Macedo Mourelle (Eds.)
Mobile Robots: The Evolutionary Approach, 2007
ISBN 978-3-540-49719-6

Vol. 51. Shengxiang Yang, Yew Soon Ong, Yaochu Jin Honda (Eds.)
Evolutionary Computation in Dynamic and Uncertain Environment, 2007
ISBN 978-3-540-49772-1

Vol. 52. Abraham Kandel, Horst Bunke, Mark Last (Eds.)
Applied Graph Theory in Computer Vision and Pattern Recognition, 2007
ISBN 978-3-540-68019-2

Vol. 53. Huajin Tang, Kay Chen Tan, Zhang Yi
Neural Networks: Computational Models and Applications, 2007
ISBN 978-3-540-69225-6

Vol. 54. Fernando G. Lobo, Cláudio F. Lima and Zbigniew Michalewicz (Eds.)
Parameter Setting in Evolutionary Algorithms, 2007
ISBN 978-3-540-69431-1

Vol. 55. Xianyi Zeng, Yi Li, Da Ruan and Ludovic Koehl (Eds.)
Computational Textile, 2007
ISBN 978-3-540-70656-4

Vol. 56. Akira Namatame, Satoshi Kurihara and Hideyuki Nakashima (Eds.)
Emergent Intelligence of Networked Agents, 2007
ISBN 978-3-540-71073-8

Vol. 57. Nadia Nedjah, Ajith Abraham and Luiza de Macedo Mourella (Eds.)
Computational Intelligence in Information Assurance and Security, 2007
ISBN 978-3-540-71077-6

Vol. 58. Jeng-Shyang Pan, Hsiang-Cheh Huang, Lakhmi C. Jain and Wai-Chi Fang (Eds.)
Intelligent Multimedia Data Hiding, 2007
ISBN 978-3-540-71168-1

Vol. 59. Andrzej P. Wierzbicki and Yoshiteru Nakamori (Eds.)
Creative Environments, 2007
ISBN 978-3-540-71466-8

Vol. 60. Vladimir G. Ivancevic and Tijana T. Ivacevic
Computational Mind: A Complex Dynamics Perspective, 2007
ISBN 978-3-540-71465-1

Vol. 61. Jacques Teller, John R. Lee and Catherine Roussey (Eds.)
Ontologies for Urban Development, 2007
ISBN 978-3-540-71975-5

Jacques Teller
John R. Lee
Catherine Roussey
(Eds.)

Ontologies for Urban Development

With 86 Figures and 2 Tables

Springer

Jacques Teller
Fonds National de la Recherche Scientifique
LEMA Université de Liège
Lab. of Architectural Methodology
1 Chemin des Chevreuils
B52/3
4000 Liège
Belgium
E-mail: jacques.teller@ulg.ac.be

Catherine Roussey
Laboratoire d'Informatique en Images et
Systèmes d'information
LIRIS UMR 5205
Université Claude Bernard Lyon
1 Bâtiment Nautibus (bat 710)
43, Boulevard du 11 Novembre
E-mail: catherine.roussey@liris.cnrs.fr

John R. Lee
EdCAAD (Architecture)
Arts, Culture and Environment
University of Edinburgh
20 Chambers Street
Edinburgh EH1 1JZ
Scotland, UK
and
Human Communication Research Centre
School of Informatics
University of Edinburgh
2 Buccleuch Place
Edinburgh EH8 9LW
Scotland, UK
E-mail: j.lee@ed.ac.uk

ISSN print edition: 1860-949X
ISSN electronic edition: 1860-9503

ISBN 978-3-642-09114-8 e-ISBN 978-3-540-71976-2

Springer is a part of Springer Science+Business Media
springer.com

Softcover reprint of the hardcover 1st edition 2007

Cover design: deblik, Berlin

Preface

Action C21 of the European programme for Cooperation in the field of Scientific and Technical Research (COST—http://www.cost.esf.org/) is dedicated to investigating *Urban ontologies for an improved communication in urban civil engineering projects.* The Action, known informally as "Towntology", brings together a large and heterogeneous grouping from across Europe, whose interests range from construction to urban tourism and from transport infrastructure to resource visualisation. On 6-7 November 2006, in Geneva, the Action convened a successful workshop to address emerging issues in the field. This volume presents the contributions to that workshop, in many cases revised afterwards to capture some of the outcomes of discussion.

Many of these contributions are from members of the Towntology group, but there are also contributions from other European researchers, and from researchers in the US. The volume represents a valuable overview of major current issues in the field of urban ontologies and encapsulates many useful and different approaches. We hope that it will serve not only as a worthy outcome of Action C21, but also as a valuable resource for a wide range of researchers.

February 2007

John Lee, Jacques Teller and Catherine Roussey
Editors

Table of Contents

Engineering Methods for Ontologies

Architecture and Construction Sector

Ontologies for an Improved Communication in Urban Development Projects

Jacques Teller

Fonds National de la Recherche Scientifique
LEMA Université de Liège, Lab. of Architectural Methodology
1 Chemin des Chevreuils, B52/3, 4000 Liège, Belgium
jacques.teller@ulg.ac.be

1 Introduction

The main objective of the COST Transport and Urban Development Action C21 is to increase the knowledge and promote the use of ontologies in the domain of urban development, in the view of facilitating the communications between information systems, stakeholders and urban specialists at a European level.

Secondary objectives of the Action are:

- producing a taxonomy of ontologies in the urban development field, contrasting existing design methodologies, techniques and production standards;
- developing an urban development ontology both in textual and visual (graph) presentation and a visual editor to integrate and update concepts, definition, photos into the ontology (software tool);
- developing a set of guidelines for the construction of urban development ontologies, based on practical examples (cases);
- analysing the role of ontologies in the daily practice of urban development.

The research work has been organized along three working groups, the first one dealing more specifically with methodologies for developing urban development ontologies, the second one dedicated to ontological issues raised by cross-comparisons between European urban development cases and the third one dedicated to practical applications of ontologies in the urban development field.

It was soon acknowledged by the COST C21 members that there is no straightforward way to define end-users' needs in terms of urban ontologies at the moment. Technology-driven approaches are not relevant as they would rapidly lead to restrict the research to the sole issue of computer representations while the ambitions of this Action extend far beyond this aspect. Furthermore conceptualizations are often tacit or implicit in the urban development domain and efforts to formalize these conceptualizations are generally viewed as "over-simplifications" by experts that are struggling to defend their scientific and technical legitimacy.

It was hence suggested to adopt a "prospective approach" in order to better identify the potential role of ontologies in fostering the exchange and support of urban knowledge. In a design-like perspective, the identification of "end-user needs" and relevant issues that could be addressed by ontologies in the urban domain should hence be considered as a product rather than a starting point of this Action. The main

J. Teller: *Ontologies for an Improved Communication in Urban Development Projects,* Studies in Computational Intelligence (SCI) **61**, 1–14 (2007)
www.springerlink.com

premises of such a prospective approach are briefly summarized in the next section, while the third section will address significant issues emerging from the work of the Action and relevant experiences in the domain of urban ontologies.

2 Prospect for Ontologies in the Urban Development Domain

Ontologies once defined as the theory of objects and their relations has certainly become a central issue in any scientific discipline, from philosophy to chemistry or social sciences. In the context of this Action, we adopted Guarino's definition of ontologies emanating from information sciences.

Guarino [1] defines an ontology as "an engineering artifact, constituted by a specific vocabulary used to describe a certain reality, plus a set of explicit assumptions regarding the intended meaning of the vocabulary words." Such ontologies are usually designed to be enshrined in computer programs. They determine what can be represented and what can be said about a given domain through the use of information techniques. Accordingly "ontology designers have to make conscious and explicit choices of what they admit as referents in a particular system or language." [2] The way to make these choices is an important subject of research given their practical implications over the long-term.

Generally speaking, the main applications of ontologies in information sciences are, on the one hand, knowledge sharing and reuse [3] and, on the other hand, the integration of data and system interoperability defined as "the ability of two or more systems or components to exchange information and to use the information that has been exchanged." [4]

In the urban development domain, both these objectives are directly relevant. Knowledge sharing and reuse is a critical issue in the view of building a common culture between experts, stakeholders and decision-makers. Interoperability between different Urban Information Systems raises issues of communication between different urban domains (cadastre, population, planning, environment etc.), scales (nation, city, district), purposes and qualities of data (2D/2.5D/3D, topologically correct/incorrect, precision).

Ontologies have also an important role to play in revealing the logical structure of existing conceptualizations. "Conceptualizations are often tacit. They are often not thematized in any systematic way. But tools can be developed to specify and to clarify the concepts involved and to establish their logical structure, and thus to render explicit the underlying taxonomy." [5] This third application may be considered as a "by-product" by specialists in ontologies. Still it appears extremely relevant in the context of this Action as urban systems have been characterized by very fast evolutions over the last decades. It is generally agreed that addressing these evolutions requires to adapt the way urban development is conceptualized. At the same time, efforts to describe the transformation of our urban systems forged a series of new concepts and neologisms (urban sprawl, emerging city, intermediate territory etc.) which partly overlap without fully covering the same reality. The relevancy of emerging conceptualizations is frequently questioned and there remains significant disagreement on the definition of key concepts commonly handled in the discipline.

3 Relevant Experiences in the Urban Development Domain

Some experiences directly relevant for the formalization of urban conceptualizations are briefly summarized in Figure 1. Even though none of these can be regarded as "plain ontologies", they inform us about difficulties inherent to our project.

	Construction sector classification	AEC Modelling	GIS Ontologies	Urban knowledge bases
Examples	ISO 12006 - 2 ISO 12006 - 3 ISO 18629 - Process Specification Language	IFC/IFG - Industry Foundation Classes	Open GIS initiative GML 3.0	EUKN - European Urban Knowledge Network COST C20 URBANET, HEREIN
Main Purpose	Standardisation Entire life cycle	Software Interoperability	Domain Interoperability	Exchange of experience Cataloguing
Leadership	Normalisation bodies	Internation Alliance for Interoperability: AEC & software industry, Public bodies	Research organisations Private agencies	European Networks Public/Private bodies
Scale	Focused on building entities (buildings, bridges) and construction complexes (motorways)	Buildings and Sites	Street networks to satellite img processing	From public spaces to urban regions
Formalism	EXPRESS bcXML taxonomy	EXPRESS ifcXML	XML, GML, OWL	Taxonomy ISO 5964 (multilingual thesauri)

Fig. 1. Relevant experiences identified in the urban development domain. These experiences are ordered from the most formalized ones *(on the left)* to the less formalized *(on the right)*

Arguably, the most formalized conceptualizations are issued from the construction sector in a view of standardization. Urban classifications tend to be less formalized but broader in their scope (ranging from heritage conservation to safety in public spaces). Further differences can be observed in the purpose of these conceptualizations, as ISO-12003 are designed for the classification of building components while Industry Foundation Classes (IFC) have been developed to allow a greater interoperability between computer models of building products.

3.1 Ontologies as a support for an improved communication

Interest in ontologies in the urban development field partly derives from the fact that communication, negotiation and argumentation are increasingly considered as essential to sound urban decision-making. Urban planning indeed evolved from pure "rationalistic models" to more transactional ones [6]. "Strategic planning", multi-stakeholder partnerships and public participation have now become mottos in the domain. Although sometimes vague in their nature and scope, the success of these notions reflects the importance of communication in present urban planning processes.

Still it has to be stressed that the so-called "communicative planning" relies on the basic assumption that stakeholders share some common understanding of terms, concepts and valid inferences, while many urban conflicts appear to be precisely fuelled by discrepancies between such basic definitions [7], [8]. Ontologies could hence be viewed as a way to address divergences between conceptual models, may it be to make these divergences more explicit and traceable.

3.1.1 Encompassing multi-stakeholder views

This was somehow the option adopted by the ISO 12006-2 standard [9] which was developed to coordinate national classifications of building products and components. The classification is intended to cover the entire life-cycle of the building from its preliminary design to its maintenance. This standard is the result of a longstanding effort of the construction sector as it was initiated in the 1950s with SfB — the first Swedish classification scheme.

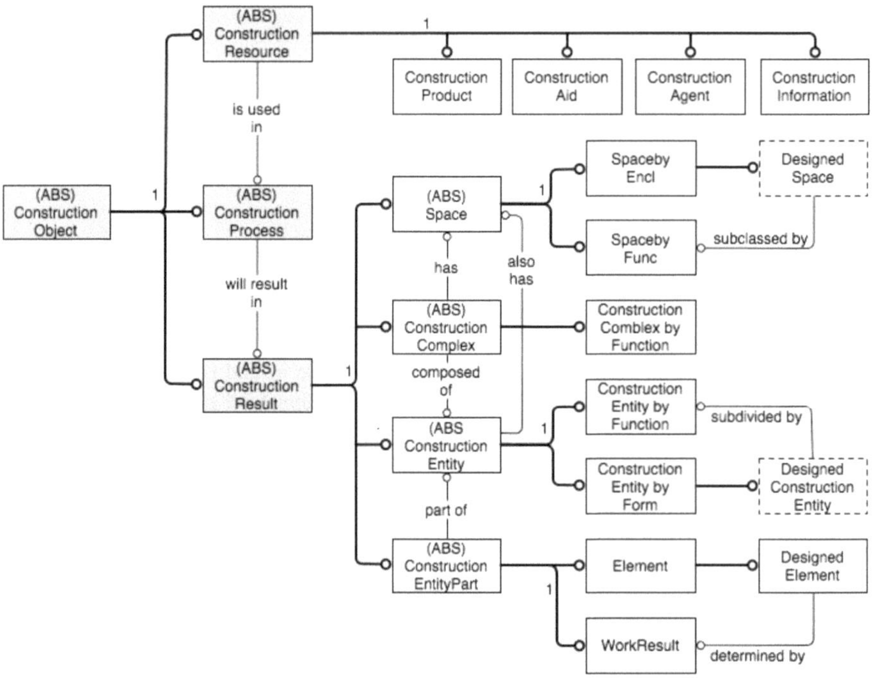

Fig. 2. The ISO 12006-2 model for classification of construction products and components, after Ekholm [10]

Quite interestingly, the ISO 12006-2 has been explicitly designed to encompass diverging views of building components. It is indeed based on three types of basic objects: construction resources, processes and results (Figure 2). The model makes a clear distinction between work results — walls or roofs for instance — and resources like products that are mobilized in the construction process — beams, bricks etc. From a conceptual point of view, a similar distinction may be established in the urban domain between mere resources (like transport systems, infrastructures) and products (such as mobility, public spaces and the like).

Construction results include construction complexes (airports, large combined buildings) and construction entities (single buildings or infrastructures) along with spaces and construction entity parts (walls, floors etc.). Most interestingly the ISO 12006-2 model defines two alternative ways to define spaces, either by their enclosure (inner space, semi-opened etc.) or their function (kitchen, living, hall etc.). A similar approach has been adopted for construction entities, as these can either be defined by their main construction method (girder bridge, arch bridge, or truss bridge) or their function-or-user activity (railroad bridge, motor vehicle bridge or pedestrian bridge). Obviously such dual views of the reality are directly relevant in the urban domain.

Even though initially designed for classification purposes, it would be tempting to use such standards in order to formalize communication between actors and thereby reduce possible misunderstandings. Still, as suggested by John Lee and Dermott McMeel in their contribution to this book, this would be oblivious of the fact that some degree of ambiguity, redundancy and even inconsistency should be admitted in communication between human actors in order to keep some adaptability to the situations at hand, to allow innovative solutions to take place (even though in an unpredicted way) and, basically, “to make urban systems work”. Sociology of action informs us of the fact that any production process can be interpreted as a chain of “translations” from initial design sketches to their progressive specification through plans, product specifications, terms of reference and final assembly. Any translation between these different steps involves a redefinition of the final object’s properties, as each of them is somehow characterized by its own “ontology”: it is now a trivial statement that a designer will not necessarily have the same ontology of buildings as a technical engineer or a construction company… The transformation of the final object through all these different translations can be formalized as a form of mapping between ontologies. This research avenue is certainly more promising than the one consisting in trying to format all communications between actors through a single ontology.

3.1.2 Support for Public Participation

Besides the above-mentioned diversity of expertise fields, public participation has now become a key communication issue in the urban development sector. Quite significantly it is now backed by significant legally-binding international agreements, as for instance the Aarhus convention, which recognizes a right of access to information and public participation in environmental matters [11]. Such conventions are applicable to the urban domain as “man-made environments” typically fall under their scope [12]. It basically means that technical information has to be made accessible to a wider audience, which may require an adaptation of interfaces and visualization tools to different user profiles and centers of interest. In this book, Claudine Métral, Gilles Falquet and Mathieu Vonlanthen suggest the use of an interface ontology in order to support a diversity of viewpoints on the same information. More significantly, the “participation revolution” implied that the general public is now increasingly viewed as a valuable provider of local urban knowledge and expertise. These authors hence propose the adoption of an ontology-based model in order to integrate and connect in the same knowledge base information coming from heterogeneous sources (Geographical Information Systems, natural language texts, personal interviews, pictures etc.), which is certainly a key challenge of present urban communication systems.

Clearly then, communication between stakeholders appears as an important application of ontologies in the urban development domain, especially when divergence about the meaning of concepts and their relations is regarded as a source of information rather than some form of pathology.

3.2 Issues of scale and versatility

One of the greatest ironies of information technology is that once conceptual structures are represented in software systems they become remarkably difficult to change, despite the inherent volatility of electronic media. In part this is because software systems are complex and require sophisticated skills and expensive resources to maintain them.

Coping with the evolution of techniques has been one of the main challenges faced by Industry Foundation Classes (IFC) since their first release in 1995. IFC classes are designed to support interoperability between building models [13]. They are now widely accepted by the industry and major Computer Aided Design software systems support IFC classes for file based exchanges with planning tools, cost evaluation applications etc.

By contrast with ISO-12006, IFC have been designed along an ad-hoc approach, without referring to an explicit model or ontology. Hence it is not clear whether the selection of building components is complete and if the classes are mutually exclusive [10]. The schema is object-oriented and proposes a deep hierarchical sub-division of building elements. Objects supported by IFC include products, processes, controls, resources, actors, groups and projects. The model was initially formalized in EXPRESS, but an XML version of IFC classes has been proposed recently. Quite interestingly IFC classes include the notion of site, which is not supported by ISO-12006-2. An IFC extension for GIS (IFG) has been developed in order to promote interoperability between Computer Aided Design software, Geographical Information Systems and urban applications like permitting systems.

A series of technical committees have been organized to support and feed extensions of IFC. One of these committees directly associates IFC designers with software companies in order to validate proposed extensions. Paradoxically such an organization further constrains possible reorganizations of the entire model, with a view to improving its overall consistency. In a somehow different approach from the one adopted by the IFC consortium, Anne-Françoise Cutting Decelle discusses the applicability of Model Design Approach (MDA) to support an increased versatility of computer systems.

MDA is based on the now "usual" idea of separating the specification of the operation of a system from the details of the way the systems uses the capabilities of its platform. Its strength resides in the mapping between different layers of computer models, from the most conceptual to platform specific, and from one version to another of the models at either of these layers. Ontologies are used to support the mapping, either for specification, abstraction or reusability and enhancement purposes. As stated by Anne-Françoise Cutting Decelle, MDA has been mostly applied in large business companies for interoperability between Enterprise Resource Planning (ERP) applications until now. It is a promising alternative to standardization approaches, in those domains like urban development where it is difficult to agree on common ontologies shared by different information systems.

3.3 Design, engineering and validation of ontologies

One of the aims of the COST C21 Action is to propose guidelines for the development of urban ontologies. A preliminary account of the state-of-the-art in the domain has been established by Roussey [14]. She distinguishes different types of ontologies according to their purpose, expressiveness and specificity. Different tools and methods to design ontologies are presented and discussed. The development process of an ontology is subdivided into six main steps: ontology specification, knowledge acquisition, conceptualization, formalization, evaluation and documentation.

Applying such general guidelines to the specific domain of urban development has been the subject of different papers gathered into this book. The proposed approaches may differ along with the method for detecting concepts, for identifying relations between these concepts and for building a taxonomy of terms.

3.3.1 Bottom-up approaches

In this book, Berdier and Roussey compare different approaches to building urban development ontologies. The first method consists in extracting concepts from technical dictionaries in the domain of road systems. The second method is based on interviews among several experts from different fields of expertise in the view of developing an urban mobility ontology. These two methods can be understood as bottom-up approaches as they are starting from the most specific concepts and tend to generalize them. Such approaches provide very specific ontologies with fine grain detailed concepts [14]. Still they may lead to problems of consistency and coherency of the ontology. Quite interestingly such bottom-up approaches may also help to reveal divergences about concept definitions and their relations, but may result in ontologies that become overtly "user-specific" with little if any possibility to be adopted by various experts/systems.

Another approach consists in extracting knowledge directly from existing databases in order to derive ontologies either through an automated process as suggested by Nogueras or through a generalization of their conceptual schema as proposed by Chaidron in this book. Nogueras applies *Formal Concept Analysis* techniques for the automatic creation of a formal urban network ontology that integrates the mappings among different road taxonomies. This allowed the integration of three local road network databases and their interoperability (SIGLA, TVIAN and AYTO). Chaidron describes the method adopted for the reengineering of Brussels' URBIS spatial databases. In a first step, the conceptualization lying behind the information system was formalized. This required a return to initial documentation and to proceed to interviews with the database managers. In a second step the authors compared the definition of concepts with the topological matrix of the

ER databases. This second step implied a further revision of some definitions in order to enlarge their scope as it helped to reveal inconsistencies in the initial ontology.

Combining these two approaches, automatic extraction of ontologies and topological matrix analysis appears as a promising avenue for deriving ontologies from urban databases in the view of their re-engineering. As urban information is more and more available in digital format, reengineering is becoming a major concern for most institutions in charge of the maintenance of these data. Data reengineering may indeed be required by the present evolution of techniques (migration from one platform to another, adoption of open-GIS format), of the requirements (new uses of the DBs, increased performance requirements, web access, inter-operability) or the data itself (integration of new information sources, 3D extensions, use of automatic acquisition techniques).

3.3.2 Top-down approaches

Two articles are addressing methods for developing top-level urban development ontologies. The benefits of top-level ontologies are that they are usually more consistent and are easy to adapt to new uses [14].

Trausan-Matu's socio-cultural ontology is based on Engeström Activity Theory and the categorization scheme of Peirce. The entire ontology is established on a basic triad that relates Subjects to Objects via mediators called Artifacts. This triad has been extended by Engeström in order to include Rules, Communities and Division of Labour. It is suggested by the author that these six top-level nodes and the relations that hold between them are capable of representing a number of diverse urban features. Actually the mere notion of Artifact, as a mediator between Subjects and Objects, but also between Rules and Communities or between Communities and Objects is certainly a stimulating one for those who are curious about the way urban objects are produced, operated and used by individuals or communities.

Finally Caglioni and Rabino propose to derive ontologies from an abstraction of urban models. After Forrester, and his seminal work *"Urban Dynamics"*, a number of models were indeed developed to better understand and predict the development of cities. Most models are addressing the relations between the development of urban economy, land use and mobility patterns (with since the 1990s a greater attention towards environmental and social issues). Caglioni and Rabino suggest that these urban models are unique sources for extracting domain ontologies as they typically include precise definition of concepts (through their inputs, outputs and main internal variables) and relations holding between these concepts (through their systems of equations). The author's demonstration is based on an ontology extraction from the Lowry model developed in the 1960s. By definition this model is based on a specific "worldview" and hence a certain domain ontology. Extracting ontologies from urban models hence appears as a way to decipher the evolution of those simplified "static models" to the complex dynamic models that are presently in use.

3.4 Ontologies for the characterization of Urban Processes

Current ontologies for information systems are mostly static, emphasizing objects with attributes and relationships over operations. They tend to minimize possible controversies about concepts, or ambiguities about their exact meaning. This may be because the roots of Geographical Information Systems are static, map-based models of the world and because of the emphasis in object-oriented approaches on attributes and relationships rather than on processes.

3.4.1 Evolution of the city's shape and limits

Quite typically, the evolution of a city's shape and its components over time are usually not encompassed in such static ontologies. This is the subject of two contributions.

The first one by Eduardo Camacho and François Golay is addressing morphological processes. The authors do not solely consider the transformation of the urban form but also the evolution of its conceptualization over time. It is generally admitted that some transformations of the urban form can not be interpreted without referring to a concomitant transformation of the way the city or some of its components were defined. Around the XVIIIth and XIXth century, the nature of the European city was largely altered as its military role literally collapsed. At the same time production activities grew very rapidly and they were more and more concentrated within urban nodes, while many cities were rather "exchange places" until then. This "ontological transformation" of the city was reflected in the urban morphology by a number of phenomena: the suppression of defence walls, the opening of large avenues for facilitating the movement of goods and people, and an unprecedented growth of the building stock to accommodate the incoming population.

It would be very difficult to interpret such morphological processes without referring to the evolution of the city's role and nature. All the more as it usually involves the "emergence" of some urban concepts. The suppression of defence walls for instance led to the creation of large "boulevards", an urban innovation that would soon be adopted in a number of urban extensions and transformations throughout Europe and that is still in use in present urban design.

Moreover, as a scientific discipline, urban morphology can be interpreted as an effort to relate the continuous development of the built environment with sporadic shifts in the way the city is conceived. The discipline actually developed in reaction against those who, in the first half of the XXth century, considered that European cities had to be almost entirely destroyed and built anew so as to cope with the demands of fast transportation systems and of the up-coming "automobile city". Research in urban morphology highlighted that, historically, radical transformations of the city's ontology never implied a total restructuring of pre-existing urban forms. On the contrary, some specific urban features like for instance medieval urban patterns or Royal places demonstrated a remarkable stability over time although the "urban systems" they formed part of had been changing radically.

In other words, the analysis of "morphological processes" should not be restricted to the evolution of the built environment but also encompass the evolution in the way a given urban feature may be conceptualized over time.

The second paper dealing with morphological processes addresses the way urban sprawl is conceptualized. Instead of using crisp delineations of urban boundaries, Hyowon Ban and Ola Alqvist suggest applying fuzzy set theory membership functions in order to discriminate between urban, exurban and rural areas. The authors argue that the definition of these areas is inherently vague and should be acknowledged as such in urban ontologies. Complimentarily the spatial implications of this vagueness can be evaluated and mapped. They demonstrate that fuzzy definitions of exurban areas are much closer to reality than crisp definitions. Still first-order logic ontology languages, like OWL for instance, do not support fuzzy memberships or fuzzy inferences at the moment. The authors suggest that these languages should hence be extended or revised in order to recognize the vagueness of some terms and to admit partial belonging to several categories. It is undoubtedly an important requirement at the time of making urban ontologies applicable to the field. Defining membership functions and measuring their spatial implications would constitute a significant step forward in the elicitation of urban conceptualisations.

3.4.2 Decision-making processes

This is all the more important as controversies about definitions can have important social, economical and political implications. Spatial processes such as sprawl or exurbanisation are intimately linked with individual and collective decision-making processes. Uncertainty not only relates to the interpretation of the 'State of the World', as exemplified by the above-mentioned case, it also applies to future decisions of individual and collective actors as well as to the likely impacts of given actions, all of which remain partly unpredictable in most cases. Plans and regulations are precisely designed to canalize anticipated investments, formalize collective intentions regarding urban development and, thereby, reduce uncertainties about the evolution of the urban system. They can hence be interpreted as a form of spatial coordination of the actions of diverse players (municipalities, urban services, private developers etc.), whose decisions are strongly interdependent. Quite paradoxically the decision-making dimension of planning is often ignored or left implicit in present spatial representations of urban development.

Lew Hopkins develops in this book a top-level ontology of urban decisions and plans. He distinguishes between two basic types of actions in terms of urban decision-making: investments and regulations. Both of which are closely intertwined and characterized by locational attributes. Decisions are defined as information about future actions. The effects of decisions are of a different nature than those of actions. He suggests to categorize urban decisions into three types: locations, alternatives and policies. Interestingly the ontology proposed by the author does not solely address the representation of 'robust' decisions and actions, but is intended to capture the net of conditional intentions from different actors that progressively shape the day-to-day urban decision-making. It certainly constitutes an important step forward in a better conceptualization of urban decision-making processes.

While the contribution of Eduardo Camacho and François Golay is addressing 'backward-looking' urban processes, the ontology proposed by Lew Hopkins is rather 'forward-looking' even though it may be used to document past decision-making processes. Urban ecology typically lies at the nexus of these two approaches as it aims to prospect local potentialities for urban development, considering the past and present states of the city, while avoiding narrow historical or geographical determinisms. As such it may be interpreted as a form of 'bridging' between both types of ontologies and certainly deserves further consideration in the view of conceptualizing urban development processes.

3.6 Language and institutional differences

The establishment of a multilingual ontology cannot correspond to the juxtaposition of N monolingual ontologies. It relies on the construction of a common conceptual taxonomy where all languages should have equal status. Still experience gained from previous attempts to build multi lingual urban development glossaries informs us of the difficulties related to this enterprise. It should indeed be acknowledged that, besides language differences, urban development conceptualizations are typically affected by their institutional context. Local development plans are for instance recognized as a key planning instrument in most European countries, but their purpose, form, content and value may somehow differ from one country or region to another.

Spanish and Italian urban planning systems are very similar to each other due to their common legal and cultural heritage, though growing differences can be observed in the nature of core instruments that form the basis of urban development practice in these two countries. Identifying differences between similar concepts may be more interesting than insisting on their main commonalities, as it fuels a critical review of the reasons and values lying behind these divergences, as well as their costs and benefits in the broader meaning.

In the same vein, Vilches and Bernabé applied the Methondology procedure to the development of urban hydrology ontologies. Quite interestingly the preliminary identification of concepts was based on the European Water Framework directive along with various other sources and dictionaries (thesaurus of UNESCO, Thesaurus GEMET etc.). This European directive provides a unified conceptual framework that has been transposed in each Member State and the proposed ontology could hence be used for inter-administrative, cross-border collaboration between Spanish and French authorities.

Such collaborations are not solely increasingly required by daily urban management issues, they tend to generalize in the view of exchange of knowledge and good practices. The European Urban Knowledge Network (EUKN) precisely aims at capitalizing and disseminating urban knowledge amongst local authorities. An e-library has been built to gather documents regarding urban policy at large (http://www.eukn.org). Quite interestingly it can be seen from figure 3 that the thesaurus designed to structure the knowledge base is very wide in scope and ambition as it spans from land use to crime prevention and integration of social groups. Arguably these different concepts are related to different "scientific disciplines" which developed their own "ontologies". Furthermore, although some documents have been translated into different languages, the taxonomy is solely available in English which is quite an important limitation given the expected audience of this library.

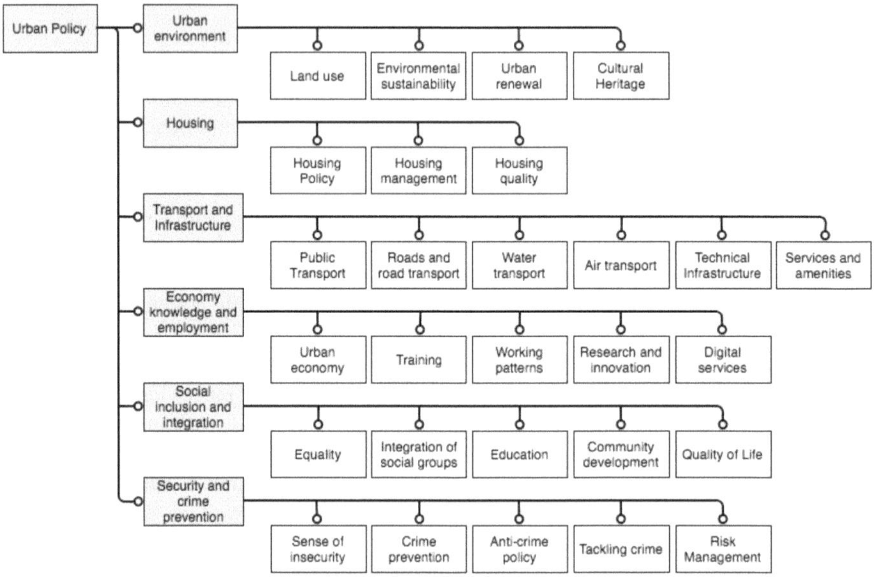

Fig. 3. The European Urban Knowledge Network (EUKN) top-level ontology. The thesaurus is composed of 254 concepts organized into five levels

Besides technical issues raised by the development and maintenance of this thesaurus, such an initiative raises challenging questions in terms of validation of knowledge included in the e-library as information comes from different fields characterized by their own authoring and review procedures, but also from local experts who may not be familiar with protocols knowledge validation. Presently the validation largely relies on National Focal Points that act as intermediaries between local users and the central network, but this issue will certainly become critical if the experience keeps growing and attracting new knowledge providers. More research is required in studying the potential role of ontologies in the view of cross-comparative analysis and evaluation of urban policies and development cases.

4 Conclusions

Even though conceptualizations are not always strongly formalized in the field of urban development, various ontologies have been developed in this domain over the last few years. Arguably some of the most "formal" ontologies emerged from the construction sector, which can probably be explained by the risks, costs and time constraints associated with this sector.

As stated in our introduction, one of the aims of this COST Action is to raise new research issues in the field of ontologies and identify their potential role in urban development. We hence deliberately included in this book references to less formal experiences, characterized by a somehow different scope than the most "established ones". Besides usual interoperability and classification purposes, novel applications of ontologies have been identified. These typically include ontologies for tracking urban decision-making processes, urban knowledge sharing and reuse at a European level or spatial database reengineering for instance.

Another objective of this Action is to progressively identify research issues that would somehow be specific to urban ontologies. Amongst these we could state the requirement to address multi-stakeholders' views and offer support for a due public participation. Versatility of concepts over time contrasted with the stability/instability of the urban form is another specific issue that probably deserves further research. Finally the urban domain has often been viewed as a battleground between different scientific disciplines (geography, history, economy, architecture etc.) characterized by divergent ontologies. This has always been a source of discussion, confusion and stimulation for those interested in its conceptualization...

Acknowledgments. The Towntology research project is supported by the European Cooperation in the field of Scientific and Technological research (COST) program of the European Science Foundation (http://www.cost.esf.org/). The number of this Action is C21.

References

1. Guarino, N.: Formal Ontology in Information Systems. Amsterdam, Berlin, Oxford: IOS Press. Tokyo, Washington, DC: IOS Press (1998).
2. Kuhn, W.: Ontologies in Support of Activities in Geographic Space. International Journal of Geographical Information Science, vol. 15 n°7. (2001) 613–631.
3. Genesereth M.R., Nilsson N.J.: Logical foundations of artificial intelligence. M.Kaufmann Pub., Los Altos(CA) (1987)
4. IEEE: IEEE Standard Computer Dictionary. Institute of Electrical and Electronics Engineers (1990).
5. Smith, B.: Ontology. In: L. Floridi (ed.): Blackwell Guide to the Philosophy of Computing and Information. Blackwell, Oxford, (2003) 155–166
6. Healey, P.: Collaborative Planning in Perspective. Planning Theory, Vol. 2, No. 2 (2003) 101–123
7. Heinich, N.: Les colonnes de Buren au Palais Royal. Ethnographie d'une affaire. Ethnologie Française, vol. 4 (1995) 525–540.
8. Zwetkoff, C.: Screening and scoping procedures. SUIT project deliverable. Available on http://www.suitproject.net/, site consulted on the 15/12/2006.
9. ISO: ISO 12006-2:2001, Building construction - Organization of information about construction works – Part 2: Framework for classification of information. Geneva: Int. Organisation for Standardization (2001).
10. Ekholm, A. ISO 12006-2 and IFC – Prerequisites for coordination of standards for classification and interoperability. ITcon Vol. 10 (2005) 275–289
11. United Nations Economic Commission for Europe: Convention on access to information, public participation in decision-making and access to justice in environmental matters. Geneva: United Nations Economic Commission for Europe, Committee on Environmental Policy (1998).
12. Teller J., Bond A.: Review of present European environmental policies and legislation involving cultural heritage. Environmental Impact Assessment Review. Vol. 22, n°6. (2002) 611–632.

13. Liebich T.: IFC 2x Edition 2. Model Implementation Guide. Version 1.6. International Alliance for Interoperability (2003).
14. Roussey C.: Guidelines to build ontologies: A bibliographic study. COST C21 Technical Report n°1. Available on http://www.towntology.net/references.php. Site accessed on the 15/12/2006.

COST Action C20
Urban Knowledge Arena: Cross-boundary Knowledge and Know-how on Complex Urban Problems

Henrik Nolmark

Managing Director
Urban Laboratory Göteborg
+46-707777255
henrik@nolmark.com

Abstract. This paper gives you an introduction to COST Action C20, Urban Knowledge Arena, and the philosophy behind the Action. New preconditions for urban development and complex urban projects are generating an increased demand for new types of competences and skills in urban knowledge. COST C20 has the objective to investigate the emerging field of integrated knowledge, experience and know-how, which is referred to as Urban Knowledge, and alongside with that, so-called Urban Knowledge Arenas. The paper also briefly describes an example of such an arena, the Urban Laboratory Gothenburg.

Keywords: urban development, urban knowledge, urban knowledge arena, trans-disciplinarity, multi-disciplinarity

1 Introduction

COST Action C20 has the objective to investigate the emerging field of integrated knowledge, experience and know-how, which is needed in today's highly complex and delicate urban development and regeneration processes. We summarize the field by using the non-established and arguable term *Urban Knowledge*. The Action, which operates during four years (2005-2009), is exploring theories, methods and tools for cross-boundary urban knowledge production, knowledge management and communication. Furthermore, we are looking at characteristics and features of *Urban Knowledge Arenas* (UKA), i.e. how to set up a platform/arena, which can give knowledge support in an urban development activity, and simultaneously feed-back into scientific research and education. Interesting examples of UKA's will be highlighted in the Action and we will draw lessons from the experience of these initiatives.

The first main focus of the Action is how the multitude of individuals, groups, professions and academic disciplines with different educational backgrounds, working cultures and traditions, working under different sets of legal frameworks and professional instruments can co-operate and collaborate in order to solve complex

H. Nolmark: *COST Action C20 Urban Knowledge Arena: Cross-boundary Knowledge and Know-how on Complex Urban Problems,* Studies in Computational Intelligence (SCI) **61**, 15–25 (2007)
www.springerlink.com

problems in the urban context. How can the diversity of professionals and scientists, including engineers, architects, sociologists, biologists, medical doctors, computer scientists, political scientists, etc, together with urban stakeholders, politicians and community groups, bring their specific knowledge, experience and know-how to form a joint platform for problem-solving and learning? As the Memorandum of Understanding (MoU) for the Action states it, C20 "...will deal with these cross-sections between scientific disciplines, different sectors in industry and public authorities and different parts of the civil society. It will take an interest in the possibilities and conditions for interlinking and cross-fertilizing knowledge of different origin, in order to develop a common ground for joint learning and mutual benefit." Consequently, these activities need a platform, a meeting place – an Urban Knowledge Arena.

2 New preconditions for urban development

Urban development has gained growing attention in politics during recent years. At the European level, the new Cohesion Policy (2007-2013) underlines the role of cities as centers of economic development. In its policy for the urban environment, the European Commission highlights the importance of cities as the places where a majority of people in Europe live, and work, but also the key role they play for the ecosystems and natural resources. This reflects the fact that across Europe, where around 80% of the population lives in cities, urban life, urban economy and urban structures have a significant impact on the well-being of Europe's citizens, as well as on biological systems. Well functioning and well managed urban areas will affect every-day life of many citizens and companies, and vice versa. Capacity building at local level becomes a key factor for improvement and a number of policy documents identify research, knowledge and learning as important elements in the strategies for sustainable urban development and social cohesion.

2.1 Dynamic and Contradictory Trends

It is clear that the systems, the relations, the actors involved and the problems in urban areas are characterized by a high level of complexity. Economy and environmental systems are becoming more and more global whereas politics and administrative systems in many parts of the world are being decentralized, and every-day life of people is getting more individualized. Demographic changes and the widespread access to international media (www, TV) are rapidly bringing new influences and lifestyles not only to metropolitan areas, but also to remote and previously isolated regions. At the same time there is a tendency that local history, identity and culture are highly praised and heritage is often regarded as an asset for local and regional development. Personal security and integrity is highly demanded, with mobility, transparency and accessibility sometimes striving in the opposite direction. Cities are indeed facing great challenges attempting to meet these dynamic and contradictory trends. In the competition for inward flow of investment capital, cities need to be attractive, exciting and innovative. At the same time there is tremendous pressure on cities to be ecologically sound and socially well-balanced.

2.2 Urban Governance: Processes, Actors and Arenas of Change

Urban development today is often undertaken by consortia through various forms of co-operation between different agents. Coupled with other changes of power relations within the political system, reference is made to a change from traditional hierarchical and sectoral government, to network dependant governance, in which influence is shared between many players, including stakeholders and citizens. The players form networks for policy and action which do not match traditional geographically based administrative and governmental boundaries. Managers, decision-makers and other key players in urban development processes need to find new roles, independent from traditional structures but clear and transparent. Management in the new types of development contexts will not be about top-down implementation of public or private policies, but a great diversity of small decisions and measures within a general framework agreed by all stakeholders. Whether you call it public-private partnerships, semi-public enterprise, public-public partnership, etc, a key issue will be to establish fruitful teamwork even though agendas may be different, and to some extent even diverging.

Furthermore, civil society has an important role to play in the development of urban landscapes. The rare and specific knowledge, which inhabitants and other local operators develop through experience and long-term commitment in a place, can be an essential contribution in any urban development process. Their often unique insights into the local context, local actors and the physical environment can be creatively coupled with general knowledge, know-how and professionalism among scientists, developers and planners. Together they can form a thorough understanding of the situation and the issues to deal with.

2.3 Complex Urban Projects

New urban landscapes and processes mean new challenges for urban professionals and decision-makers. Innovative urban development and regeneration has to take into account the existing situation, together with political agendas, the needs of inhabitants and business actors into visions for the city, which can be realized and manageable in the long run. This means that a stronger integration is needed between the different scales and time-frames of planning and management. Long-term strategic planning is far from "touching the ground" if it becomes obsolete and needs to interact with urban design, technical implementation of projects and the every-day management of the city. This means that any urban project of some scale today is carried out in a complex organizational and political environment, and there is a tremendous pressure on actors involved to work and act in a cross-sectoral, cross-organizational manner.

Still, each department or expert in a specific field utilizes criteria, methods, procedures and vocabulary specific to the discipline or organization it belongs to.

These criteria and methods are often unfamiliar to the experts in other sectors or disciplines involved in the same project, or process. The consequence is often a series of uncoordinated actions producing a sub-optimized result, far from reaching overarching societal goals.

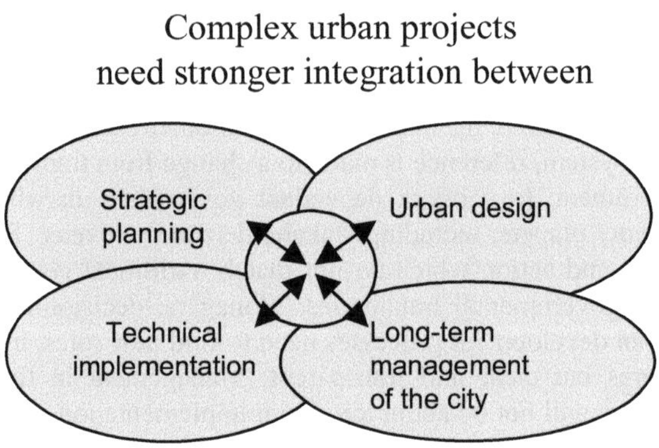

Fig. 1. Urban project

3 Urban Knowledge

In recent years, we have seen several promising pilot-projects with the aim to establish and develop cross-boundary co-operation among professionals and experts with different backgrounds, also involving researchers and communities. Even though the ultimate goal is to produce a new physical urban environment, or to reach socio-economic regeneration, these activities can in some ways be regarded as urban knowledge arenas, where knowledge and experience is exchanged between the different groups and individuals, and new knowledge and ideas are being generated.

We are definitely in a transition from sectoral and disciplinary working cultures towards more comprehensive and cross-disciplinary ways for learning and knowledge production. However, it is often hard to overcome difficulties due to historical and cultural barriers that have grown among the different disciplines and sectors over centuries. Aims and efforts have been primarily geared towards finding optimal solutions to various single issues and problems, with less attention put to the holistic approach.

Co-operation among experts and personnel from different technical departments (e.g. urban planning and transportation) is quite common today, whereas professional and scientific cooperation between technical and non-technical departments or organizations is still less developed. In addition to that comes the difficulties in communication between the non-expert community groups and citizens, and in particular groups who are not familiar with terminology and culture in such environments.

3.1 Urban Knowledge – Cross-boundary Knowledge and Know-how for Urban Issues

The complex urban issues, and the context in which they are dealt with, are generating an increased demand for a deeper research-based urban knowledge, both in terms of disciplinary research and in cross-boundary production, management and implementation of knowledge. However, in order to make a real improvement of the situation, there is a need to find new ways, corresponding better to the needs of the people responsible for urban development action.

Let us make it clear that Urban Knowledge is a non-established term, and it may also be contested. It is not a new discipline or a new field of science. It might be better thought of as a way of drawing together, refining and managing knowledge, which is often generated in a wide variety of sources, representing a multitude of diverse disciplines, theoretical and methodological traditions as well as from practical experience and lay judgment. It can also be referred to as a way of learning and implementing knowledge and know-how, in a situation of action in the urban environment (see fig. 2).

Many of the solutions occupy the interfaces between different sectors and scientific fields and a fellowship is needed between humanists, social scientists, medical scientists, natural scientists, architects, planners, engineers and so on, together with stakeholders and engaged citizens. The future development of urban knowledge will take place through recurrent meetings and joint projects involving all those who share the interest in applying their specific knowledge and experience to a joint knowledge-building enterprise.

Experience from previous attempts in the same direction, among them the UK initiatives to form a research program on Urban Sustainability in the 1990's show that interdisciplinary cooperation is called for and often a necessity, however very difficult to organize and implement fully in line with the intentions[1]. The conditions for such work thus need to be illustrated and thoroughly analyzed.

3.2 Current Situation in the Production and Dissemination of Urban Knowledge in Europe

Urban development, as stated above, is often characterized by its conflicting nature, between contradictory external trends and driving forces, and contradictory interests according to social and economic positions among those involved and/or affected by the situation. The success of development action can in this sense be regarded as the capacity of a community to cope with contradictions, between different complex objectives, and turn the differences in opinion and interest into a creative process.

Correspondingly, the capacity of a scientific society to deal with the complex problems which occur in urban development and regeneration is related to the ability to manage the diversity of the sometimes contradictory, and even conflicting disciplinary perspectives and patterns of thinking and turn it into a synergetic knowledge, in which the outcome is more than the sum of its disciplinary

[1] Marvin & Evans

components. This implies the ability of the project or the organizational structures to open up the compartmentalization of scientific knowledge and the sectoral division of responsibilities in contemporary academic organizations without losing relevance and quality in relation to the fundamental scientific requirements.

There are several interesting tendencies and examples of interdisciplinary, multidisciplinary and trans-disciplinary initiatives for urban knowledge and know-how. Many of these are action-oriented and make linkages not only across scientific boundaries but also between research and professional practice. Bridging between research and policy-making also becomes more and more evident, as reflected in the current interest in bridging the *Science-society* gap.

The field of *Urban Design* has its point of gravity in architecture and has a long tradition of collaborating with urban engineering, sociology and sometimes also other types of expertise such as environmental psychology. *Urban Engineering* originates from civil and environmental engineering fields, but is opening up towards social and behavioral science. In urban design and urban engineering, there is a strong tradition of cooperation between practice and academic work. The school of *Urban Studies* is based in socio-economic analytical research (e.g. political science, economy, geography, sociology), but is currently in a phase of opening up across the scientific spectrum. The international community in *Urban Ecology*, often strong in systems analysis, has the ambition to broaden the field to include more economic and social aspects, moving towards viewing cities as socio-ecological systems. Both urban studies and urban ecology have close relationships to policymaking.

At the European level, there are a few interesting initiatives in recent years, among them the Fifth European Union Framework Program for Research (FP5), Key Action *The City of Tomorrow and the Cultural Heritage* is perhaps the single largest and most influential. The COST domain *Urban Civil Engineering* has mobilized large groups of European scientists in its bold and ambitious attempt to integrate "process thinking", i.e. planning organization, decision-making etc with "product thinking", i.e. engineering, design, construction etc. The European Science Foundation *Forward Look on Urban Science* aimed to be an urban science policy program to guide research programs. Correspondingly, in 2004 the European Commission, DG Environment commissioned a *Working Group on Urban Environment Research and Training Needs*, as part of the preparatory work for the EC *Thematic Strategy on the Urban Environment*. In its final report, the group recommended a "code of urban research", in which it is stated that "research needs to be inter-institutional, and inter- and trans-disciplinary", and that "(topic-) relevant groups and disciplines need to be included".

Other interesting initiatives are the UNESCO MOST project, the European Urban Knowledge Network, the Metropolis network project, the Millennium Assessment project and many other initiatives at international, national, regional and local level. Altogether, these activities form an emerging movement towards more sophisticated types of integrative, interlinking and cross-fertilizing production of knowledge for complex urban development and regeneration situations. The growing tendency to summon people of different backgrounds, with the objective of producing knowledge for action in complex fields, is sometimes referred to as "boundary organizations". The structural support in terms of international associations, programs and platforms to meet and strengthen the networks is still rather weak, and one of the aims of COST Action C20 is to contribute to filling this gap at the European level.

4 COST Action C20[2] – Urban Knowledge Arena

COST C20 has the general objective, as stated in the MoU "to explore and develop a European arena for cross-boundary, integrated knowledge and know-how on complex urban problems..." In doing that, the Action is exploring possibilities to bring the different approaches (urban design, urban engineering, urban studies, urban ecology etc) one step further into an *Urban Knowledge School,* by exploring and developing the novelty in theories, methods and tools and by analyzing and characterizing different innovative types of organization, management and implementation in relation to urban knowledge. This refers to the notion of an Urban Knowledge Arena, which can operate at local as well as regional, national, European or international level.

4.1 Work Program

The work program consists of three main elements:

- WG 1 Characterization of existing and future urban knowledge: to identify, explore and exchange experience on theories, methods and tools, which can facilitate urban knowledge. The WG tries to uncover the meaning behind the keywords *urban knowledge arena* and how to construct procedures for an integrative process, which can benefit from the involvement of various forms of knowledge, in order to make better solutions for our cities.
- WG 2 Facilitating a European arena for urban knowledge: the WG has moved from the original purpose, as expressed by the title of the WG, into a discussion on how to set up a knowledge arena, characteristics, and success/failure factors integrative process, which can benefit from the involvement of various forms of knowledge arenas in the participating countries.
- WG 3 Innovative initiatives in research, policy, practice: to identify, give feed-back and disseminate innovative initiatives in urban research, urban leadership and urban projects.

The intention is that the work will result in a conceptual book on urban knowledge as the scientific output of the action, a summary report with messages to policy makers,

[2] COST (European Cooperation in the field of Scientific and Technical Research is an instrument to support co-operation among scientists and researchers across Europe. COST operates through specific thematic Actions, which generally runs for a period of 4 years. COST can be regarded as a networking activity, enabling exchange of knowledge and experience at a European level. A COST Action should not be understood as a research project.

including priorities for further urban research. In addition to that there will be individual papers and workshop reports/proceedings[3].

4.2 The First Year of Operation

From this point of departure, as there is no international state-of-the-art as such in the field, a lot of effort during the first year of operation has been spent in the meetings on comparing and discussing different conceptual understanding of the terminology related to urban knowledge, and to reach a common understanding on concepts of knowledge production, management, dissemination and making use of knowledge in urban development situations. A number of working papers have been written addressing the questions of knowledge in an urban context. Several presentations have been given, e.g. experiences of urban research initiatives in the UK, France and Hungary, international state-of-the-art in theory and sociology of science regarding different modes for production of knowledge and knowledge management. Complementary to the conceptual and theoretical discussions, three workshops have been carried out, with thematic focus on urban issues, illustrated by local case studies. The objective of workshops in C20 is to investigate the role urban knowledge has had in these cases of urban development, thereby feeding into the continuous discussion and analysis in the Working Groups of the Action. The 1st C20 Workshop was based on the case of the high-speed railway station in Liège, Belgium, and highlighted the relation between an international, high-profile development investment and the capacity of the city to integrate it into the local context and in its long-term development strategies. The 2nd workshop was held in Oeiras, Portugal, and focused on the relation between territorial development and socio-economic policy. The third workshop focused on community involvement and public participation in urban development processes, illustrated by the case of "Dialogue Southern River Bank" in Gothenburg, Sweden.

4.3 Results so far

The results so far are a) the consolidation of the C20 network, with representatives of 21 European countries and a junior network with 10 members, b) a higher level of awareness and common understanding on some concepts involved in urban knowledge and urban knowledge arenas, and shared experience from some examples of research programmes and research-policy-practice cooperation, c) a preliminary conceptual model for characterization and analysis of urban knowledge arenas. The next year will be focused on the analysis of interesting examples of Urban Knowledge Arenas and the continuous discussion on issues such as; the governance and power of knowledge, integrated urban knowledge – what is it, when is it relevant and for whom, what will be the consequences, what strategies are developed in relation to the utilization of different forms of knowledge etc.

---

[3] Visit the web site of the Action for further information on papers, reports etc, http://infogeo.unige.ch/uka/

5 Example of an Urban Knowledge Arena: The Urban Laboratory Gothenburg, ULG

The Urban Laboratory Göteborg (Gothenburg), ULG, was formed recently as a joint venture between the City of Gothenburg and Chalmers University of technology. The aim is to be able to bridge the current gaps between academia, professional practice, policy and civil society in the field of urban development. It is regarded as a platform for innovative cooperation, integrating knowledge and know-how from the different partner organizations into urban action in Gothenburg. ULG provides a framework for a large network of people with different backgrounds, doers as well as thinkers, to meet for mutual learning, joint generation of knowledge, exchange of experience and supportive action in the world of urban change and development.

ULG is politically independent, non-for-profit and it is run in partnership, regulated through a contract. Both partners contribute financially and in-kind to keep the platform running for a first period of four years, 2005-2009. Each project/activity carried out on the platform is managed and financed through a separate contract. Although the two partners manage and coordinate the platform, ULG is open to other organizations to participate in activities. Recent and current activities include:

- Local Action: Contribution to the organization and setting-up of a public dialogue process, the Dialogue Southern River Bank
- Local Action: Contributing to the organization of an Urban Safety Workshop in the urban district *Kvillestaden* (together with Young Urban Network and Master students at Chalmers University of Technology)
- University: Development of new concepts for master level education at Chalmers School of Architecture, *Urban design and development*
- University: Development of a PhD course, *Tools and processes for sustainable urban development*
- Connecting Gothenburg to the international community in urban knowledge: Co-organizing events at UN-HABITAT World Urban Forum 3 (Vancouver 2006)
- Connecting Gothenburg: Coordinating COST C20, Urban Knowledge Arena
- Connecting Gothenburg: Coordinating a project on "Cultural actors, methods, tools and expressions in urban development and governance processes", together with national and local actors in arts, culture and town planning.

The Urban Laboratory Göteborg is developing a concept for local urban knowledge arenas, in which education and training, local capacity building, R&D, debate etc will be carried out in relation to current real-world urban projects and processes in Gothenburg. By forming these UKAs, the ULG is hoping to realize synergies between the different types of activities and a better basis for generation, management and dissemination of knowledge related to the particular urban districts, or sites.

References

1. COST Action C20 Urban Knowledge Arena, Memorandum of Understanding, MoU. (the MoU is an official COST Document, adopted by the COST Committee of Senior Officials, regulating the Action at a general level. The MoU is written by a group of contributing authors, including C. Cardia, A. Dupagne, T. Kleven, T. Muir, H. Nolmark, R. Nolmark and K. Strömberg)
2. Researching the sustainable city: three modes of interdisciplinarity (S. Marvin & B. Evans, Environment and Planning A 2006, vol 38)
3. The New Production of Knowledge: The Dynamics of Science and Research in Contemporary Societies (Gibbons et al., Sage 1994)
4. Re-thinking Science (Nowotny et al., Sage 2001)

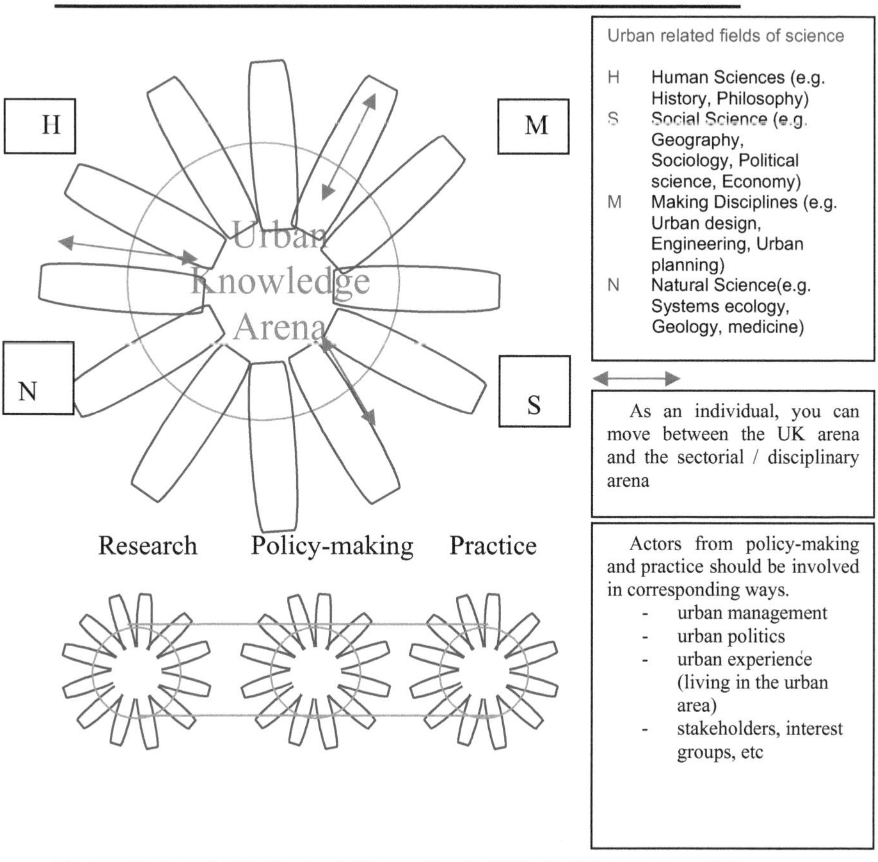

Fig. 2. Urban Knowledge Arena (from COST C20 MoU)

Pre-consensus Ontologies and Urban Databases

Robert Laurini

LIRIS - INSA de Lyon, F - 69621 Villeurbanne
Robert.Laurini@insa-lyon.fr

Abstract. Facing the difficulties of interoperability and cooperation between several urban databases, a solution is based on ontologies which can help not only clarify the vocabulary used in urban planning, but also organize urban applications; indeed multiple definitions can be given to various urban objects. So this is the scope of the Towntology project which aims at defining ontologies for urban planning whose design is characterized by the multiplicity of definitions. After having presented some ways of using ontologies for various actors in urban applications, a definition of pre-consensus ontologies is given, together with some groupware tools to collect multiple textual and multimedia definitions in sub-ontologies, to check and consolidate the vocabulary in order to reach some consensus. We conclude this paper by giving some recommendations for the Towntology project for covering the whole urban field by integrating various sub-ontologies.

Keywords: Urban ontologies, Towntology project, consenus, pre-consensus ontologies, groupware

1 Introduction

One of the main problems we have to face in urban information systems is the problems of interoperability and cooperation between several databases [8]: indeed, each database was created independently from others, i.e. with different entities and attributes with different meanings. Usually, two levels of interoperability are defined, the lower level called syntactic, and the upper level called semantic. As the syntactic level is solved through OpenGIS[1] recommendations, a general solution of the semantic level is based on ontologies in order to deal with the meaning of vocabulary. But in reality, the vocabulary problem is not only a database problem, but more important is the clarification of the vocabulary used by all actors dealing with urban databases, and especially by urban planners.

For this reason, the Towntology project was launched in 2003 at INSA Lyon in order to create an urban ontology between urban planners and computer scientists (see [7] or [9] for details). Then facing the difficulty to cover the whole urban field, a COST group[2] was created and placed under the responsibility of Jacques Teller [10]. Now, it regroups more than 15 laboratories in Europe.

---

[1] Refer to http://www.opengeospatial.org/standards.

[2] Refer to http://www.towntology.net/.

R. Laurini: *Pre-consensus Ontologies and Urban Databases,* Studies in Computational Intelligence (SCI) **61**, 27–36 (2007)
www.springerlink.com

The scope of this paper is to give an overview of problems we have to face in order to define urban sub-ontologies and to integrate them into an unique domain ontology.

This paper is organized as follows. In the second section, we will address the necessity of ontologies in urban applications, and second the organization of groupware to create urban ontologies.

2 Necessity of ontologies for urban applications

In this section, examples for interoperability in urban applications will be detailed in order to show how ontologies can be used to solve those problems.

2.1 Examples of interoperability

The main examples of interoperability in urban databases can be seen in street repairs in which different databases can be used, not only belonging to the municipality (sewerage, traffic light control) but also belonging to different companies such as for water supply, electricity, gas. Other examples can be found in environmental assessment (for instance dealing with pollution control of an international river such as the Rhine or the Danube), and for providing new pervasive services (Location-Based Services).

Let us examine an example in the cooperation of several urban databases, linked to physical hypermedia [2]: find the roadmap for going from the Da Vinci Gioconda painting in the Paris Louvre Museum, to the Velasquez Meninas painting in Madrid Prado Museum. The solution must be found by means of the cooperation of several databases:

- from the Louvre database for exiting from the Gioconda to the next metro station,
- from the Paris Transportation Company to go from the nearest metro station to Paris Airport,
- from the Airlines database to fly from Paris Airport to Madrid Airport,
- from the Madrid Transportation Company for going from the airport to the nearest metro station,
- from Prado database for going from the nearest metro station to the Meninas painting.

2.2 Definition of ontologies

The word "ontology" comes from Greek "*Οντος*", Being and "*Λογια*", Discourse, so meaning the discourse about existing things. More precisely, ontology refers to the theory of objects and of their relations. Gruber [5] defines an ontology as "*an explicit specification of a conceptualization*", and Guarino [6] "*An ontology is an engineering artifact, constituted by a specific vocabulary used to describe a certain reality, plus a set of explicit assumptions regarding the intended meaning of the vocabulary words*". An important aspect is that the various actors must agree about the definition of objects and their relations; so we speak about ontological commitment between actors.

Pragmatically, a common ontology defines the vocabulary with which queries and assertions are exchanged among actors. Ontological commitments are agreements to use the shared vocabulary in a coherent and consistent manner. From a computing point of view, an ontology can be seen as a semantic network.

But in the case of urban planning, there exist many different definitions of key-objects such as "city" or "road".

In the Wikipedia[3], one can find a dozen of definitions of the word "city", but none addresses the whole urban complexity. After Toynbee [11], a city can be defined as a human habitat which cannot provide all food they need, whereas other defines a city as petrified expression of power structures [3]. How to combine those definitions into a single expression?

Let us consider another problem regarding the definitions of "streets". Let us consider three actors in the same city, street cleaners, postmen and gas men: they all can claim "*we do have a street file*". In reality those files are slightly different:

- street cleaners only clean public streets, so their file only is composed of public streets;
- in theory postmen passes in all streets, but when a cul-de-sac has letter-boxes in a main street, they do not consider those cul-de-sac streets
- for gas men, their file only consists only in streets in which residents have gas.

As a conclusion, even if the concept of street can receive an acceptable definition from urban planners, analyzing several databases can reveal that definitions are different. Generalizing this, we can claim that in practice, even if two databases are using the same word (street), the probability is high that there exist some hidden differences in the definitions.

In other words, multiplicity of definitions is often hidden behind similar terms. To solve this problem, one solution is to define contextual ontologies (See [1] for details).

2.3 Ontology-based interoperability

To ensure interoperability, one way is to use ontologies. In the framework, each database is assigned its own local ontology perhaps written from its conceptual model. Moreover a domain ontology is used as a sort of bridge between both local ontologies (i.e. linked to a database) as illustrated in Figure 1. By means of those ontologies, a mediator is generated made in two parts, one for translating the initial query to be accepted by the second database, and the second to transform the results (See Figure 2).

Let us take a small example in demography, with two databases, (i) DB1 with one entity `residents`, and (ii) DB2 with two entities `men` and `women`. How can we get the number of `men` and `women` separately in DB1, and the total number of `residents` in DB2? The second case can be solved by an exact mediator, so giving: `DB2.residents= DB2.men + DB2.women`. However, for the first case, only approximate mediators can be generated, for instance:

---

[3] Refer to http://en.wikipedia.org/wiki/City.

```
DB1.men = 0.48×DB1.residents
DB1.women = 0.52×DB1.residents.
```

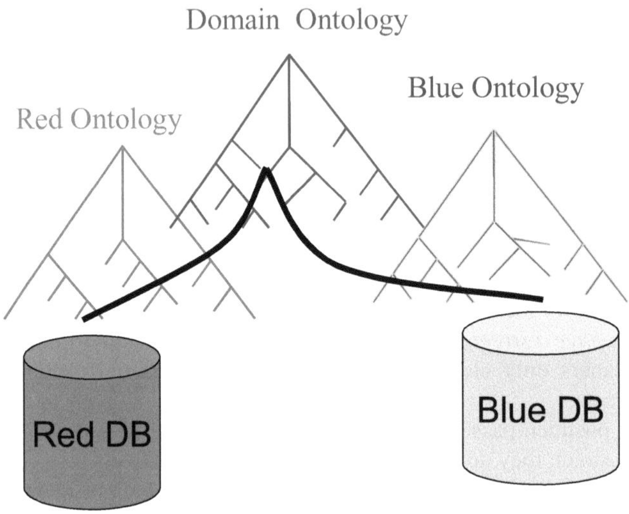

Fig. 1. Using domain ontology to ensure interoperability between two databases

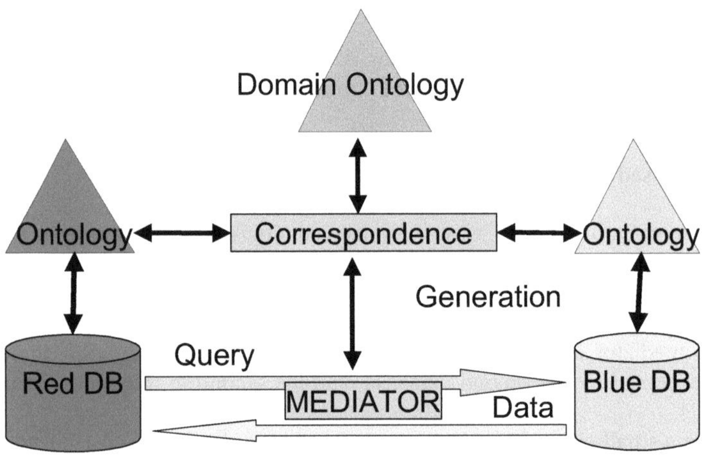

Fig. 2. Generation of mediator to ensure interoperability

The previous formulae can be included into the query-part of the mediators. For the transformation of the results, let us mention an example in distances with different units. For instance the data-part of the mediator can transform distances in kilometers by distances in miles.

From a language point of view, ontologies are generally written with languages such as OWL[4] which derives from description logics.

2.4 Specifications of pre-consensus urban ontologies

Ontologies are easy to define in applications where the vocabulary is well standardized [4]. The topical example is chemistry.

As illustrated in the previous paragraphs, in urban planning, the context is totally different especially due to the variety of definitions. So before translating some textual definition into OWL, some consensus between actors must be found. Now, we can introduce two kinds of ontologies, pre-consensus and post-consensus ontologies as depicted Figure 3. As the majority of existing ontologies can be considered as post-consensus, in our case, our domain ontology in urban planning is a pre-consensus ontology whose main characteristics is the necessity of a repository to collect existing definitions. Then, when all definitions are accumulated, actors can convene to look for a consensus; and when the consensus is reached, translation into OWL can start. It is important not to forget cultural and linguistic problems in this task.

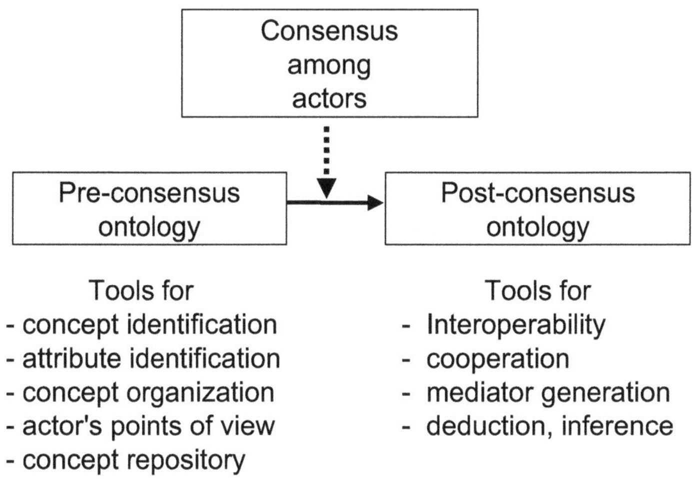

Fig. 3. Differences between pre-consensus ontology, and post-consensus ontology

So, a repository must be design to collect multiple definitions and attributes. Since in some cases multimedia definitions must be considered, for instance for noise definitions or in architecture when drawings and sketches are necessary. Another important issue is lineage and traceability of definitions. Finally not only a repository must be defined, but also software tools to manage the various pre-consensus ontologies together with adapted human visual interfaces. Figure 4 illustrates those

[4] Refer to http://www.w3.org/TR/owl-features/.

visual various access methods, (i) from graphs of concepts (semantic network), (ii) from photos illustrating various concepts. In addition to those visual methods, a third one was added based on the alphabetic list of concepts. Since the ontology is represented as a graph, a nice visual solution is to access directly from the graph and to navigate from concepts to concepts. Another interesting access method can be based on photos in which several zones can be activated, especially zones representing concepts; in other words, several photos of cities can be used as entry point into ontologies.

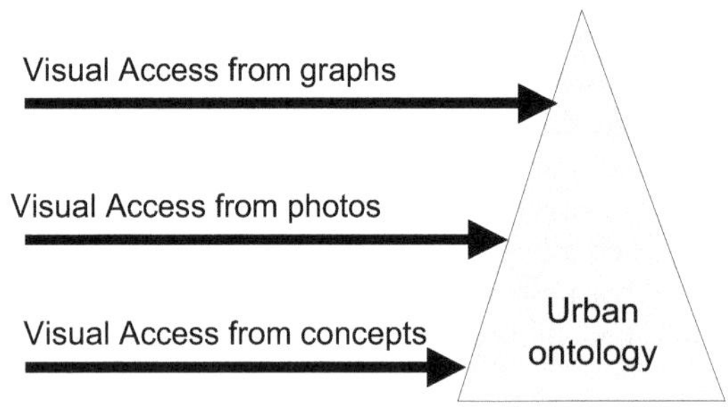

Fig. 4. Various ways of accessing to the ontologies.

The access methods based on photos is very important for us because the user is very familiar with urban scenes as given in photos. For integrating a photo into the system, one needs to find rectangular zones corresponding to concepts. Let's take the example of a rectangular zone surrounding a bus. According to the level of abstraction, this zone can correspond to several concepts:

- bus itself,
- bus as a mean of public transportation,
- public transportation,
- transportation of passengers,
- etc.

Finally, the main characteristics of our system are as follows:

- Semantic network,
- Hypertext structure,
  - Multiple definitions,
  - Origin and lineage of definitions,
  - Possibility of updating,
  - Photos and drawings
  - Visual presentation.

All main objects of our pre-consensus ontology can be regrouped into a conceptual model given Figure 5.

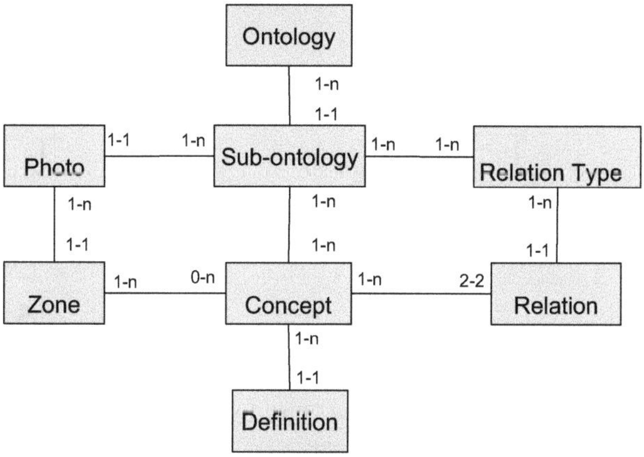

Fig. 5. Conceptual model of a pre-consensus ontology

3 Groupware for pre-consensus urban ontologies

As said earlier, the scope of the Towntology project is to define a complete urban ontology. For that, each laboratory interested is developing its small sub-ontology. The role of the groupware system will be to help those laboratories define pre-consensus sub-ontologies, i.e. collect the various multimedia definitions including lineage. For that, each group of actors can work independently on the definition of their important terms. In other words, they need frequently to add some fresh definitions or update them into the repository. When a sub-ontology is ready, it will be presented to the groupware system which will integrate it. Of course a sub-ontology can refer a concept already present in another sub-ontology.

After having very rapidly presented the description language, the groupware system will be sketched.

3.1 Language

Since OWL was not adequate to our problem, taking all those aspects into consideration, a new language was created to store all multimedia definitions into our repository. This language is an extension of XML, some excerpts of the structure of which are given Figure 6. The main divisions are `HEAD` and `BODY`. `HEAD` regroups some identification and metadata regarding this ontology, whereas `BODY` is really the core of the ontology: the reader can see that any concept can have various multimedia definitions, and every update can be traceable.

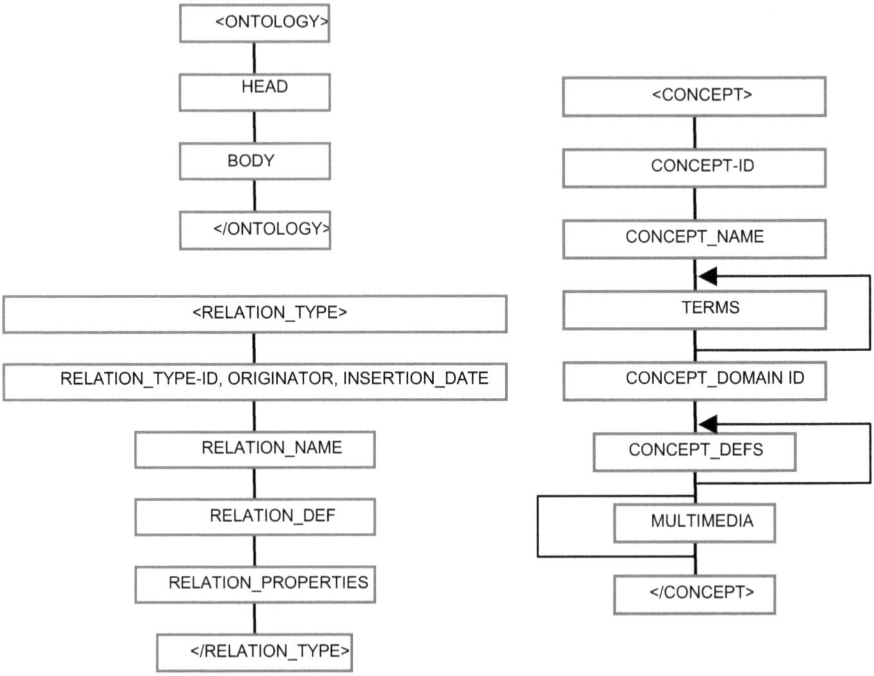

Fig. 6. Exerpts of the structure of the Towntology Language

3.2 Groupware system

The groupware system consists of two parts, the first one for creating sub-ontologies, and the second one for integrating sub-ontologies.

For the creating of ontologies, based on the previous language, three modules were written:

- navigating and browsing a sub-ontology, essentially based on the three types of accesses as illustrated Figure 4,
- updating a sub-ontology, especially by adding new concepts, new definitions and new multimedia resources,
- and preparing an image that can be used as an entry into the sub-ontology; mainly this image is split into rectangular zones which addresses one or several concepts.

The second system is for integrating a new sub-ontology. It consists in several modules:

- validating the proposed sub-ontology, essentially by checking the grammar and some integrity constraints,
- and validating cross-references of concepts with two main aspects: a relation can refer a concept located in another sub-ontology, or a new definition can be added to a concept already stored elsewhere.

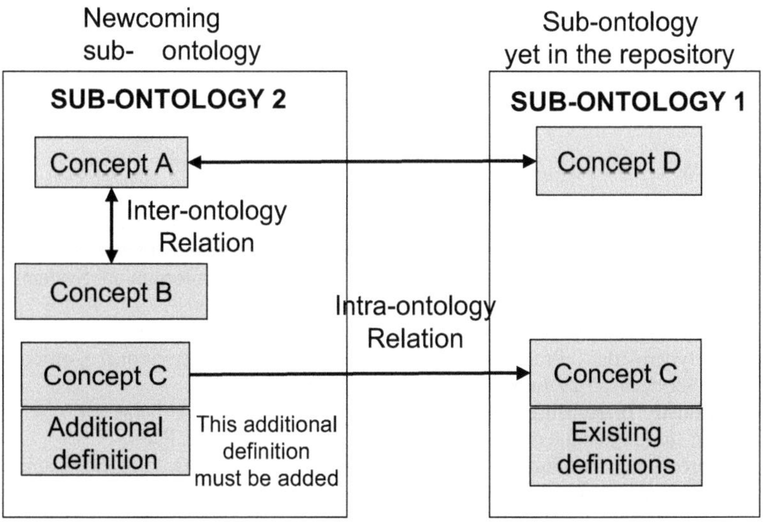

Fig. 7. Integrating a new sub-ontology into the ontology repository

When all those tasks are made, the new sub-ontology is integrated into the system.

Another module must be written for the seamless browsing of the sub-ontologies. Until now, when a sub-ontology refers to a concept which is located in another ontology, the system only show it; indeed, it could be of interest to continue the browsing without taking those divisions into account.

Several examples of sub-ontologies will be found in this book, together with examples of the navigating system.

4 Final remarks

To conclude this paper, let me remind that the Towntology project was initiated with two scopes, (i) interoperability of urban databases, and (ii) clarification of the vocabulary used by urban planners. Presently, a first software tool exists that can be used to define sub-ontologies. For me the first goal is to populate the various sub-ontologies in order to cover the whole urban field, and for that a first tool was created.

When all definitions are collected, the next step is to consolidate those sub-ontologies and check completeness. The subsequent phase will to reach to some consensus; a special tool must be defined, and also a tool for helping the users transform those verbal or multimedia definitions into descriptive logics, so as to code with OWL.

Concerning the language problems, this is not so easy. A naïve way could be to collect terms and definitions in English, and after to translate everything into other languages. One of the first difficulties could be to translate all definitions of very important terms such as cities, towns, urban planning, etc., terms which have sometimes no EXACT counterparts in other languages, especially due to historical, cultural aspects. Similarly a difficulty arises for the translating of legislative terms.

References

1. Benslimane D., Arara A., Falquet G., Maamar M.Z., Thiran P., Gargouri F. (2006). "Contextual Ontologies: Motivations, Challenges, and Solutions". In Proceedings of the Fourth Biennial International Conference on Advances in Information Systems 18-20 October, 2006 Izmir, Turkey. ADVIS 2006 (Springer), 2006.
2. Challiol C., Gordillo S., Rossi G., Laurini R. (2006) "Designing Pervasive Services for Physical Hypermedia". Proceedings of the ICPS'06: IEEE International Conference on Pervasive Services, 26-29 June 2006, Lyon, France, pp. 265-268.
3. French Ministry of Housing (1977) "Programme de Recherche Urbaine du VIIème plan, Paragraphe sur la Poétique des Villes". Ministère de l'Equipement, Paris, 1977.
4. Gómez-Pérez A., Fernáez-López M., Corcho O. (2004). "Ontological Engineering" Springer Verlag Advanced Information and Knowledge Processing, 2004.
5. Gruber, T.R. (1993) "A Translation Approach to Portable Ontology Specifications". In Knowledge Acquisition, 1993, Vol. 5, N° 2, pp. 199-220.
6. Guarino N. (1998) "Formal Ontology and Information Systems", Amended version of a paper appeared in N. Guarino (ed.), Formal Ontology in Information Systems, Proceedings of FOIS'98, Trento, Italy, 6-8 June 1998, Amsterdam, IOS Press, pp. 3-15 URL: http://www.ladseb.pd.cnr.it/infor/Ontology/Papers/FOIS98.pdf
7. Keita A., Laurini R., Roussey C., Zimmerman M. (2004) "Towards an Ontology for Urban Planning: The Towntology Project". In CD-ROM Proceedings of the 24th UDMS Symposium, Chioggia, October 27-29, 2004, pp. 12.I.1.
8. Laurini R. (2001) "Information Systems for Urban Planning: A Hypermedia Cooperative Approach", Taylor and Francis, 308 p. February 2001.
9. Roussey C., Laurini R., Beaulieu C., Tardy Y., Zimmermann M. (2004) "Le projet Towntology : Un retour d'expØrience pour laconstruction d'une ontologie urbaine". "Revue Internationale de Géomatique", vol. 14, 2, 2004, pp. 217-237.
10. Teller J., Keita A., Roussey C., Laurini R. (2005) "Urban Ontologies for an Improved Communication in Urban Civil Engineering Projects". Int'l Conf. on Spatial Analysis and GEOmatics, (SAGEO 2005), 2005 Avignon, France, June, 20-23, 2005.
11. Toynbee AJ (1970) "Cities on the Move". Oxford University Press. June 1970.

How Formal Ontology can help Civil Engineers

Stefano Borgo

Laboratory for Applied Ontology, ISTC-CNR, Trento (IT)
borgo@loa-cnr.it
www.loa-cnr.it

1 Introduction

In this paper we report some considerations on the developing relationship between the area of formal ontology and that of urban development. Even in the studies on urban and territorial systems we register a phenomenon common to most applied domains: the increasing interest on ontology and the difficulties to understand its novelty. Indeed, the area of applied ontology spans a variety of methods and ideas, some of which have been developed much earlier in other approaches. This older group of 'ontological tools' (among which we find classification methods, taxonomic organization, graph and lattice theories) are well-known techniques and form the basis of most university programs (from engineering to geography, from computer science to cognitive science). It is natural that the domain experts that want to introduce applied ontology to their domain find easy to get hold of these old techniques since, in a sense, these are already part of their background. Unfortunately, these techniques have already reached their limits and now have little to say in ontology research:[1] they are substantially the same as thirty or forty years ago (even relatively recent proposals like dynamic taxonomies are just innovative applications of well-known knowledge techniques).

In contrast, it is harder for non-ontologists to understand the new ideas and techniques that applied ontology has to offer since they often are obtained by mixing ideas from disparate field like philosophy, region-based geometry and logic. This fact is not surprising because ontology is a recent and innovative area of research which has not found a proper place in education programs yet. A few compelling aspects can be identified: ontological research aims at general principles and rules which make it more abstract than the previous approaches to knowledge representation (consider the conceptual shift from the discussion of 'data' to that of 'entity' or even 'possible entity'). It applies subtle distinctions imported from the philosophical domain (like substance vs accident, tropes vs properties) which are new in conceptual modeling. Furthermore, it concentrates on good and deep formalizations of the adopted concepts (thus breaking away from the limits of conceptual systems). The combination of these and other elements explain the novelty of applied ontology and the problems it has to be properly understood by practitioners.

---

[1] This claim does not want to contrast their usefulness which is even higher today essentially for the improvement of modern informatics systems. They are valuable tools and are successfully applied in many situations. Nonetheless, they are of less interest (since not innovative) in ontology research as developed from the late 90s.

S. Borgo: *How Formal Ontology can help Civil Engineers,* Studies in Computational Intelligence (SCI) **61**, 37–45 (2007)
www.springerlink.com

In what follows, we address some (and somehow scattered) issues of interest to civil engineers, architects, and experts in urban development that are sensitive, on the one hand, to the theoretical foundations of their domain area and, on the other hand, to improve the stability and reusability of their models via ontological techniques. However, before we can introduce these issues (the problem of incompatible space representations, consistent use of linguistic resources, integration of existing and disparate domain ontologies) we need to set some basic distinctions that serve us to put some order on the class of ontological systems. After all, we need to agree on what we mean by 'ontology' if we want to consistently compare alternative views and arguments.

2 Classifying Ontologies

Ontology systems (or simply ontologies) are complex systems that can be analyzed from a variety of perspectives: language, content, taxonomic structure, domain coverage, semantics and so on. Each perspective provides a different way to classify ontologies. Here it suffices to look at two of them, namely, the semantics and expressivity of the adopted language and the generality of the included concepts.

The first classification gives us a way to classify ontologies according to the language and the type of semantics it adopts. This is a crucial distinction: ontological systems are not simple classification structures, they are supposed to classify entities according to their *essential nature*. We can capture it in the ontology only through a careful use and interpretation of the adopted language. Since the major tool we have to ensure the correct interpretation of the language is formal semantics, it is important to know in which *semantic class* the ontology is positioned. Here we identify three general classes. The first includes the systems with the weakest semantics (in terms of formal semantics) since they necessarily rely on natural language. This class collects mainly linguistic and terminological ontologies, comprising the vast majority of ontologies today available. A second class includes systems usually limited to weak formal languages. The main concerns in developing ontologies in this second class are related to complexity, feasibility, and other implementation issues (which affect the generality of these systems). In the third class we find quite expressive logical theories with full formal semantics.

Once we have the semantic classes available, we can look at the formal expressivity of each system (the formal distinctions that the system can consistently make) to refine the classification. (Note that the subclasses provided here are not exhaustive.)

1. **Linguistic/Terminological ontologies**
   [these are ontologies committed primarily to the semantics of natural languages]
   - Glossary
   - Controlled vocabulary
   - Taxonomy
   - Thesaurus
2. **Implementation driven ontologies**
   [in these systems the primitives are committed to natural language semantics and the derived terms to formal semantics]

- Conceptual Schema
- Knowledge Base

3. **Formal ontologies**
   [these ontologies commit exclusively to the semantics of formal languages]
   (types are given by classes of interpreted languages like modal, predicative logics, logics with binary relations only, logics with restricted models, etc.)

The other classification instrumental to our goals is independent from the above and looks at the concepts the ontology uses to categorize entities. Such a classification is harder to provide since content is hard to define. Fortunately, for our goals it suffices to consider a rough and general classification regarding primarily the broadness of the concepts included in the systems (see also [1]).

1. Domain ontologies
2. Core (reference) ontologies
3. Foundational ontologies

2.1 Formal ontologies: the notion

Most people rely on a widely cited description of ontology which says: "An ontology is an explicit specification of a conceptualization." ([2], Sect.2). We think that the general acceptance of this notion is due in large part to the lack of constraints it puts; any collection of terms, graph of classes, and logical theory can be seen as an ontology according to the above notion. Nonetheless, Gruber's proposal gives an important intuition on what an ontology is. Then, it is important to find a technical definition that correctly separates proper ontological systems from others.

Formal ontology [3] explicates and deepens Gruber's intuition. Guarino's proposal is to add specific constraints in order to avoid misinterpretation (and misuse) of the system. In his view, an ontology must be based on:

I) a set of basic linguistic elements and a set of precise rules to construct terms and relations (adoption of a *formal* language)
II) a clearly stated semantics for the language (adoption of a *formal* semantics)
III) a rich set of explicit motivations and arguments, possibly with references to the philosophical and ontological literature, to justify and illustrate the adopted categories and relations (presence of documented *philosophical* analysis)

The above requirements constrain the technical aspects of an ontology without affecting the content. This choice makes clear that applied ontology is a scientific domain that looks at the *formal* properties of the entities it studies, i.e., the ontological systems. Regarding the content, condition III) sets a minimal request: it requires it to be well documented. No restriction is put on the view the ontology professes since this aspect is what determines its acceptance as a knowledge representation tool, not its quality as an ontological system.

With the above definition of formal ontology, it becomes possible to split the complexity of standard knowledge representation systems into two distinct parts that, by and large, correspond to the ontological component and the knowledge-base component. The first, which is the domain of formal ontologies, deals with the organization of the knowledge structure while the latter is concerned with the information contained in the knowledge structure.

2.2 An example we are all familiar with: MATH

We all have been exposed to mathematics and understand the basics. The isolation of the mathematics domain and the precision of its objects and techniques make this science suitable for challenging our intuition on what ontology is.[2] The classification of page 2 suggests that we may give several different answers. An analysis of the proposed ontologies for maths helps us since it allows u to make explicit the position we take in this paper. The reader should try to write down its own answer and compare it with the one we give below.

First, recall that mathematics is a specific language formed by terms, sentences, function symbols, quantifiers, etc. which is used to talk about special entities like sets (e.g. $\emptyset$), numbers (e.g. π), ordinals (e.g. $\aleph_0$), functions (e.g. log_e), matrices (e.g. $\left[\begin{smallmatrix} 0 & 2 \\ 3 & 3 \end{smallmatrix}\right]$) etc.[3] The entities are individuated via primitives (which come together with an axiomatization) and definitions (derived notions).

Everyone would accept that neither a language, nor a collection of entities is *per se* an ontology. This observation holds as well for the language of mathematics and the set of its entities. We continue that the collection of primitives and derived notions of mathematics (let them be concepts or relations) is tantamount not an ontology. Indeed, from the perspective embraced in this paper, we conclude that the ontology of mathematics is the *complex structure of relationships connecting primitives (as concets) and derived notions.*

2.3 What is a (formal) ontology then?

Leaving aside the variety of things people mean when using the term 'ontological system' or 'ontology' for short (a labeled graph, a set of terms, a knowledge base, a structure for knowledge etc.), one must recognize that there is a clear-cut distinction between a system *for knowledge organization* and a system *of knowledge.*

As we said, ontologies are developed to cover the first of these two senses, i.e., they are systems developed to organize knowledge. More than that, the success of the term 'ontology' is due, in our view, to its explicative import which is realized only when the system is coupled with a description of the view on the 'world' (or domain of interest) that has motivated it. Unfortunately, some researchers minimize this aspect and claim that the ontology structure itself suffices as an (implicit) description of the ontology viewpoint. Then, they do not feel committed to go further in analyzing the ontological aspects purported by the system. Most systems in the class of terminological ontologies are a consequence of this 'permissive' reading of the notion of ontology. Others

[2] Clearly, we posit the question from the perspective of applied ontology. The ontology of mathematics from the perspective of the philosophy of math is a different (although related) issue.

[3] Of course, in all these examples we refer to the denotations of the listed terms or expressions.

work with weak languages in which one cannot formalize even quite basic constraints. This is the source of another important fault of several systems: insufficient (actual) formalization.

We think that proper ontologies must address two main aspects:

- the *structural* aspect: the system clearly establishes and describes the types of existing entities, the structural organization and relationships among the types
- the *formal* aspect: the system is constrained with a sufficiently rich axiomatization that rules out (most) possibilities of misinterpretation

2.4 ...and what is a foundational ontology?

Foundational ontologies are formal ontologies that provide a structure for the most general types of entities. They characterize the meaning of general terms like entity, event, process, spatial and temporal location (as opposed to drilling machine, driving, being in London, the 2004 olympics) and basic relations like parthood, participation, dependence, and constitution (as opposed to mechanical parthood, playing a card game, depending on water, having an arm).

The purpose of foundational ontologyies is abstracted away from any direct application concern. These systems aim to provide a formal description of entity types and relationships that are common to *all domains* and to provide a consistent and unifying view of 'reality' from a given perspective. In principle, any (consistent) ontology is justified by a foundational ontology, i.e., by a general view on what exists and how (ontological) classes of things are related.

3 The DOLCE ontology

DOLCE [4] stands for the Descriptive Ontology for Linguistic and Cognitive Engineering. It is a foundational ontology that concentrates on *particulars*, that is, roughly speaking, objects (both physical and abstract), events, and qualities. It does not classify properties and relations: these are included in the system as far as needed to characterize particulars. DOLCE adopts the distinction between objects (like houses and refrigerators) and events (like cutting and visiting) and differentiates among individual qualities, quality types, quality spaces, and quality values as we will see. Technically, it is a formal ontology that relies on a very expressive language, first-order modal logic.

DOLCE adopts a *multiplicative approach* since it assumes that different entities can be co-localized in the same space-time. For example, a building and the amount of matter that constitutes it are captured in DOLCE as two distinct entities (as opposed to different aspects of the same entity). The reason lies on the different set of properties that these entities enjoy: the building ceases to exist if it collapses due to a earthquake since a radical change of shape occurs while the amount of matter is not affected (the identity of an amount of matter is not affected by the change of the shape). For a different example (discussed at length in the philosophical literature), consider a statue made of clay. DOLCE models the statue and the clay as different entities which share the same spatial (and possibly temporal) location. This allows us to capture the strong intuition that a scratched statue has changed (since scratched) and yet it is the same statue it was before. In DOLCE these claims are consistent since the statue itself might not be affected by (minor) scratches, but the clay (which is the constituent entity of the statue) does because amounts of matter cannot loose parts.

The category of *endurant* collects entities like a "railroad" or material like "some cement", while events like "making a hole" and "driving a car" are in the category of *perdurant*. The term 'object' itself is used in the ontology to capture a notion of unity or wholeness as suggested by the partition of the class "physical endurant" into the classes "amount of matter" (whose elements are (an amount of) gold, air etc.); "feature" (a hole, a corner); and "physical objects" (a building, a human body). See Figure 1. Note that the terminology adopted departs sometimes from the usage in the knowledge representation area since it has been affected in part by philosophical literature.

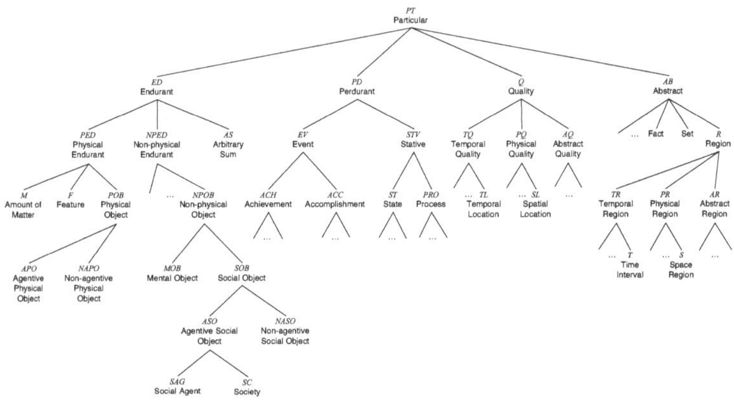

Fig. 1. Taxonomy of DOLCE basic categories (from [4])

Both endurants and perdurants are associated with a bunch of *qualities*. These entities and their evaluation are crucial in DOLCE and the distinction between *individual qualities*, *qualia*, and *quality spaces* has been set with the aim of capturing common sense in a coherent and consistent way as we are going to see.

Qualities and Incompatible Representations

Adopting the DOLCE perspective, one can set a framework [5] where different forms of quality representations can coexist and consistently (as well as coherently) interrelate. The basic entities, as mentioned above, are *individual qualities*, e.g. the weight of this brick. Individuals qualities inhere in specific individuals so that the weight of this brick is different from the weight of that brick, no matter how similar they are. Furthermore, individual qualities can change through time since the weight of this brick matches 2 kg now and will match 1.9 kg after I cut off a corner piece. *Qualia*, e.g. a specific weight, form another type of entities. These entities are obtained by abstracting all possible individual qualities from time and from their hosts. Then, differently from individual qualities, qualia are not entity dependent. Nonetheless, analogously to individual qualities, qualia are divided in types: weight qualia, shape qualia, color qualia, and so on. If two bricks put straight the pivot of a perfect balance, then they have the same weight quale although they have different individual weight qualities. In this sense qualia represent perfect and objective similarity between (aspects of) objects. Finally, *spaces* corresponds to different ways of organizing qualia. They are motivated by subjective (context dependent, qualitative, applicative, etc.) similarity between (aspects of) objects. By means of spaces, a structure can be imposed on qualia (for example ordering, metrics, geometry and qualitative relations) and this makes it possible to differentiate several quantitative and qualitative degrees of similarity. With these distinctions it becomes possible to talk about the weight of a building in different ways as indicated by the first column of Fig. 2. Analogously, for the other qualities.

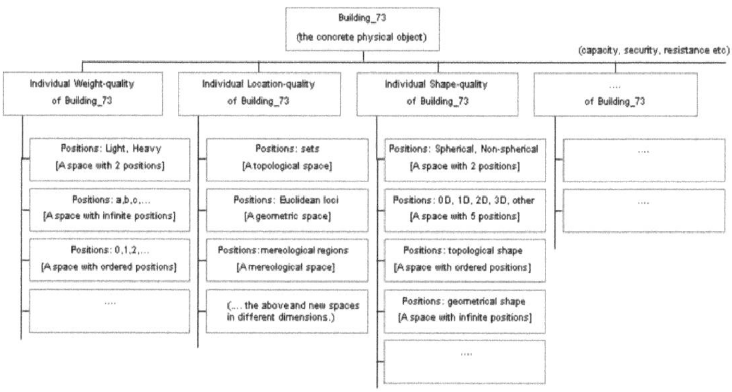

Fig. 2. Space Modularity in DOLCE

This modularization techniques allows the use of different space representations within the same ontology. Indeed, location in DOLCE is simply an individual quality that physical entities must possess. The comparison of the location of one object with the location of another is carried out consistently in any space of interest (or even across spaces) as suggested by column 2 of Fig. 2.

4 Coupling Foundational and Weaker Ontologies

Even the optimistics would admit that it will take many years before a rich formal ontology that covers the urban development domain can be available. Also, one may doubt that such a system is needed in practice. The solution might be to find a good balance between the time- and resource-consuming effort that the development of a reliable formal ontology requires and the inexpensive and prompt availability of terminological domain ontologies. Fortunately, the adoption of a foundational ontology already suffices to greatly improve the robustness and interoperability of existing (implementation driven or even terminological) domain ontologies. From this observation, what is necessary is a careful extension of the foundational ontology with appropriate concepts that correctly organize the main categories to which core and domain concepts can be connected. This view brings forward the interesting problem of coupling foundational and weaker ontologies. The analysis of the problem (including the study of proposed solutions which in the literature are mostly based on the WordNet linguistic resource [6]) shows that different techniques can be applied.

There are basically four major strategies [7]:

1) *Re-structuring*. The ontology is used at the meta-level only. The real focus is an ontological improvement of the linguistic resource that does not require the addition of ontological categories or relations. In particular, the computational properties of the linguistic resource are unaffected.
2) *Populating*. The ontological and linguistic systems are here treated as simple taxonomies. The focus is on the mapping between these two taxonomies. The map is then used to enrich the ontology with lexical information.
3) *Aligning*. In this case the focus is on both the ontology's structure level and the linguistic object level. This approach consists in implementing both the previous perspectives of re-structuring and populating. The result, which cannot be reduced to any of the original systems, is ontologically sound and linguistically motivated.
4) *Merging*. The first step consists in isolating a system that takes the common parts of the ontology and the lexical resource. Then, the system is extended (by choosing among the alternative views given by the original systems) to ensure enough coverage. The approach relies on techniques for redundancy removal and consistency preservation.

5 Appling a Foundational Ontology

A final remark is in order: foundational ontologies *are* implementable. However, even if a foundational ontology is fully implemented, it cannot be used in the same way as terminological ontologies. The two types of systems have different roles [3] as we mentioned earlier.

The DOLCE foundational ontology is available in first-order modal logic and has several versions in different languages[4] like KIF, OWL-DL, DAML+OIL and RDFS. The *Common Algebraic Specification Language* (CASL), developed by The Common Framework Initiative [8], has been enriched with an extension, HETS, to manage foundational ontologies and their modularization; the full DOLCE ontology (including a partial modularization) is now available in the CASL system as shown in [9]. In particular, the possibility to manipulate ontologies as modular systems is crucial when dealing with large logical theories like DOLCE. Indeed, the special approach of CASL to ontology construction borrows from research in logical studies and software engineering, and is driven by applicative concerns. As a result, in a system like CASL, it becomes possible to store several domain ontologies and to reliably transfer information from one another provided they are linked to a common foundational ontology like DOLCE.

[4] For further information, visit http://www.loa-cnr.it/DOLCE.html

At the cost of complicating the system, one can even adopt different foundational ontologies, each connected to a group of domain ontologies, and transfer (part of the) available information through ontological systems that embrace very different views on 'reality'.

Acknowledgement

Most of the work reported in this paper is the result of research developed at the Laboratory for Applied Ontology (LOA) in Trento. The author has been supported by the Provincia Autonoma di Trento and the national project TOCAI.IT.

References

1. Borgo, S., Gangemi, A.: At the core of core ontologies. In Gangemi, A., Borgo, S., eds.: Proceedings of the Workshop on Core Ontologies in Ontology Engineering, CEUR-WS.org/Vol-118 (2004) 1–4
2. Gruber, T.R.: Toward principles for the design of ontologies used for knowledge sharing. International Journal of Human-Computer Studies **43** (1995) 907–928
3. Guarino, N.: Formal ontology in information systems. In Guarino, N., ed.: Proceedings of the Second International Conference on Formal Ontology in Information Systems, IOS Press (1998) 3–15
4. Masolo, C., Borgo, S., Gangemi, A., Guarino, N., Oltramari, A.: Ontology Library (Wonder-Web Deliverable D18). Available at http://wonderweb.semanticweb.org/deliverables/documents/D18.pdf (2003)
5. Masolo, C., Borgo, S.: Qualities in formal ontology. In Hitzler, P., Lutz, C., Stumme, G., eds.: Foundational Aspects of Ontologies (FOnt), Universitat Koblenz (2005) 2–16
6. Fellbaum, C.e.: WordNet An Electronic Lexical Database. Bradford Book (1998)
7. Prevot, L., Borgo, S., Oltramari, A.: Interfacing ontologies and lexical resources. In: Ontologies and Lexical Resources: IJCNLP-05 Workshop. (2005) 1–12
8. CoFI: CASL Reference Manual. LNCS 2960 (IFIP Series). Springer (2004)
9. Luettich, K., Mossakowski, T.: Specification of ontologies in casl. In Varzi, A., Vieu, L., eds.: Proceedings of the Thrird International Conference FOIS 2004. Volume 114., IOS Press (2004) 140–150

Ontology for Land Development Decisions and Plans

Nikhil Kaza and Lewis D. Hopkins

Department of Urban & Regional Planning
University of Illinois at Urbana Champaign
111 Temple Buell Hall, 611 Taft Drive
Champaign, IL, 61820, USA
nkaza@uiuc.edu l-hopkins@uiuc.edu

Abstract. The focus of geographic and other ontologies of urban development has been to represent locations with object attributes or objects with locational attributes. Urban information systems should also represent decisions, which have or could have locational attributes. Development processes are critically influenced by expectations about declared intentions manifest through plans and records of decisions. This paper provides an ontology of decision situations characterized by actors participating, intentions expressed, and alternatives considered. We argue that these elements are closely tied to and interdependent with other aspects of urban ontologies, which typically focus on physical objects of development. An ontology of plans and decisions will enable sharing of information among actors and consideration of disparate and distributed information.
Keywords: Plans, Decisions, Urban Ontology

1 Introduction

Representations of urban development have focused on spatial objects over time and attendant functional relationships [1]. Important components of urban development processes, however, include intentional actors who plan for their own actions and respond to decisions and plans that are made explicit by others. A city is not only a physical entity, but also an *institutional entity.* This paper develops an ontology of the actors, decision situations, and plans that make up the institutional structure of a city. This institutional ontology is essential in order to represent considerations and strategies for providing urban infrastructure.

Hopkins [2] argues that two types of actions are crucial in planning: Investments and Regulations. Investments are changes in assets. Regulations are changes in capabilities of actors more specifically in rights. Plans are statements of intentions about how these investments will be made, at least in the sense of some level of implied commitment. Urban planning is concerned with the choices of actions (or combinations of actions) situated in a spatio-temporal context and intended in relation to goals. Laurini [3] describes some of the approaches to operationalise planning documents in urban information systems.

N. Kaza and L.D. Hopkins: *Ontology for Land Development Decisions and Plans,* Studies in Computational Intelligence (SCI) **61**, 47–59 (2007)
www.springerlink.com

When the definitions of choices, goals, and actions are broadly construed, planning is about intentions, decisions taken prior to action, and possible 'sequences' or otherwise related sets of actions. Plans are records of such decisions, including their intentions and recognized relationships among actions. This theory of planning is consistent with the theory by Bratman [4] who argues that intentions are predicated upon plans and vice versa. However, in order to keep track of intentions of others as well as our own, we need an ontology that is general enough to be useful and extendible enough to apply to many different legal and other institutional contexts [5].

Planning, by the State or otherwise, requires that plans consider the nature of interdependence of our own planned actions on others' plans and the evolving set of circumstances. To plan effectively, one must be cognisant of information regarding the following questions. 1) What is the 'State of the World'? 2) What institutional framework (rights, regulations, and norms) permits certain kinds of actions and prohibits others? 3) What are the intentions of other players in the process? 4) How are changes to the institutional framework fashioned? 5) What changes to the state of the world are implied by changes to assets and regulations? Relevant answers to these questions are needed in a system that could support making plans, and using plans. The questions become interesting because of the issues of space, time, interdependence, and contingency in land use planning.

In this paper, we argue that representing decisions in urban planning ontologies is important from the urban planning perspective. Decisions raise expectations, provide indications of commitment, and are typically precursors to actions that change some aspect of the world. Our approach is different from "Argumaps", which represents arguments with spatial attributes as described in [6] or [3]. Argumaps are useful to chart various alternative arguments and positions of interested stakeholders that are tied to specific locations thereby aiming to reconcile them. In contrast, representing decisions and attendant decision situations helps in discovering alternative as well as contingent decisions when the decision making capacity and authority are distributed.

2 Decisions, Actions, and Effects

Elsewhere, in [5], we have described an ontological framework for representing urban development processes. The purpose of this paper is to elaborate the descriptions of the decisions of the intentional actors who populate urban systems and the relationships between them. A decision situation is characterised by the actor or a collection of actors who are deciding, alternatives considered, and plans that inform it. A decision situation may or may not result in an explicit decision. When the decisions are being made, recognition of interdependence with other decisions is informed by the plans (Fig. 1). The actors have the capability to make such a decision, specifically the decision is within the jurisdiction of the actor. A decision situation considers alternative actions and chooses a subset of these to be pursued at a future date. Plans help in decision situations by pro-

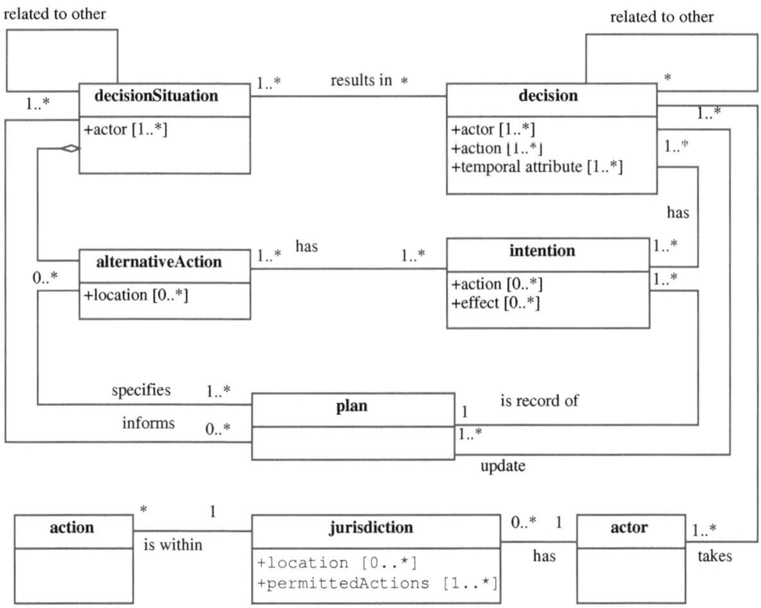

Fig. 1. Plans & Decisions

viding information about the framework within which the decision fits and by addressing the questions of irreversibility and interdependence with other decisions and actions. Plans, as records of intentions, are continually modified, when new information about these dependencies emerge. Figure 2 describes the concept of realised actions in relation to decisions. These actions have certain effects, intended and unintended. By making these decisions explicit, especially through adopted urban development plans or by any other such public proclamations, the actors shape the expectations of other actors, which influence other decisions and actions. It is thus important to sort out the differences between decisions and actions. We characterize decisions as information about intended actions. A decision to build a road is different from building the road. The increase of the traffic flow on the road and the resultant rise in the property values of adjacent properties are effects of building the road. However, speculative development may raise the property value of the adjacent property even before the road is built, merely as an effect of publicly announcing, in some credible fashion, the decision to build the road [7].

It is to identify these distinct effects that we distinguish decisions from actions. Using Bratman's terminology, a decision is an explicit 'volitional commitment'[4]. While actions change the state of the world, decisions merely provide information about how these actions are situated in the future. These decisions may not result in realisation of any of the chosen actions, changes in the polices,

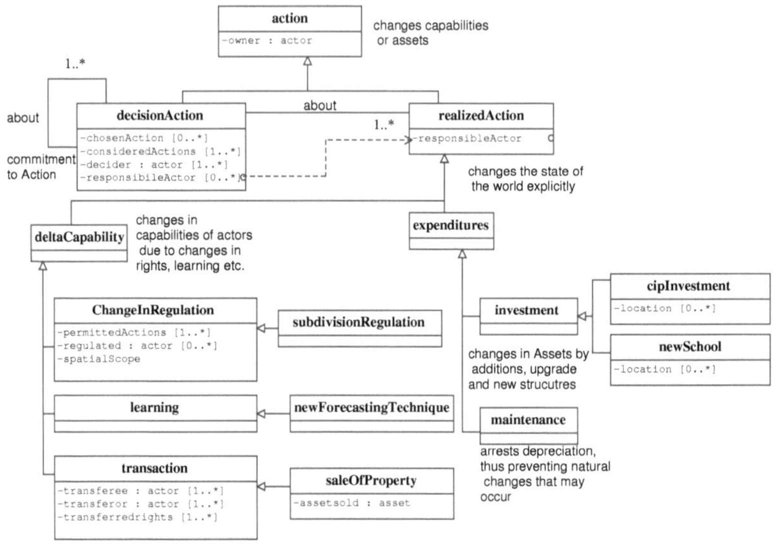

Fig. 2. Action Class Diagram – Adapted from [5]

or any other changes in the 'state of the world'. The mere fact that these decisions are taken provides indications about commitment, thereby generating expectations and thus influencing other's decisions. This section's title may convey some sense, perhaps slightly misleading, of a sequential relationship of decisions, actions, and effects. A decision to act need not result in the actual action, and the action may not realise its original intentions; the relationships between them are more circuitous and thus richer.

3 Types of Decisions: Location, Alternatives, and Policies

Figure 3 explicitly illustrates distinctions between different types of decisions that are manifest within plans. Numerous other examples of these types of decisions as well as other types can be found in plans.

A decision that has a spatial attribute is illustrated in Fig. 3(a). The realignment of the existing Olympian Drive north of Urbana to a new location is marked in the plan. While the new road has not yet been built, and the existing road is still in use, it is useful to have information about the decision, the intent to realign this road, which will also lead to a sequence of other decisions. At the time of the publishing of this plan, the intended alignment, represented by the dashed line ends at Illinois Route 45, and the decision to terminate it at that junction or to continue it to High Cross Road has been deferred to a later date.

Figure 3(b) illustrates another type of information in plans, a restricted set of alternatives for which a decision is not yet made. The exact location of the

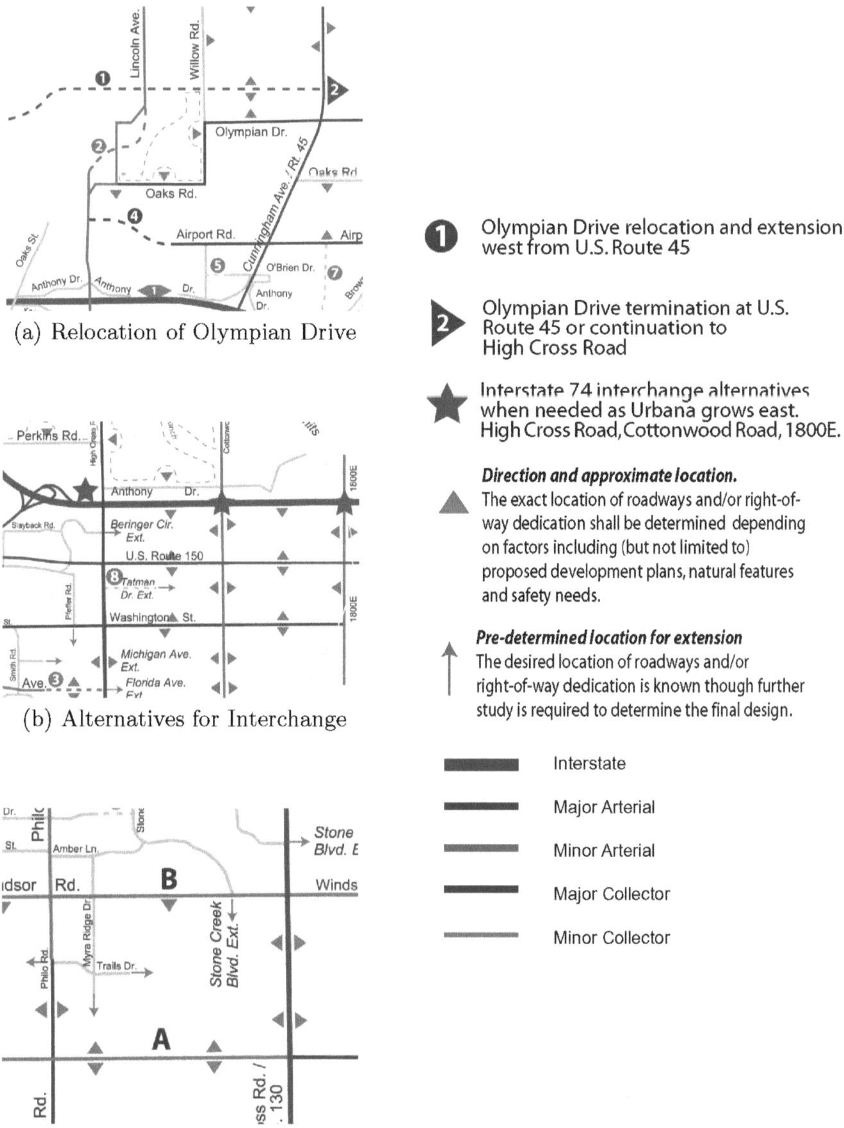

(a) Relocation of Olympian Drive

(b) Alternatives for Interchange

(c) Policies for Sub-collector Streets

Fig. 3. Types of Decisions – Excerpted from [8]

interchange on Interstate 74 has been a contentious issue for the City of Urbana and the neighbouring residents. While the location of the interchange has not been determined, three alternatives (represented by three stars at High Cross Road, Cottonwood Road, and 1800E) have been identified. It would appear from the information in the plan that the future decision on where the interchange will occur will consider only these three alternatives or slight modifications of these.

Figure 3(c) illustrates a policy specification as information in a plan. The policy is that two sub-collector streets should be built between two existing parallel collector streets (**A** and **B** in the figure), which are generally 1.6 kilometres apart. While the exact locations of the rights of way of these sub-collectors were ambiguous at the time the plan was published, the policy is nevertheless very pertinent information about the city's intent about infrastructure investments. The triangles are intended as 'sliders' indicating the need to identify the end points and connecting alignments when other decisions about land subdivision are made in the future. The intersection node may be specified, however, if a sub-collector already exists on the other side of **A** or **B**, which then fixes the location of the endpoint in the interest of continuity. This situation is represented by a different kind of arrowhead as shown in the legend.

These examples demonstrate, though not exhaustively, different types of decisions that are made and their implications for making inferences about changes in the physical state of the world. They demonstrate different sets of alternatives considered and chosen, and different kinds of information about intentions before resulting action. While not all decisions are made explicit in publicly available plans, most government decisions have to be made public in some way prior to initiation of actions. These examples also point to the spatial, topological, and temporal relationships among decisions, which result in similar kinds of relationships among the actions and effects.

3.1 Location

Hopkins et al. [5] argue that none of the attributes of urban development are fundamentally tied to a location. In particular, it is reasonably obvious that decisions themselves may not have locations as attributes. The actions that are a part of the decisions may have locations. The actor's jurisdiction may have a spatial attribute. This divorcing of the location from the ontology of the urban processes is important. Planning information systems have long relied on the intellectual development of geographic information science which is fundamentally focused on spatial objects, but planning requires a different frame [9].

Assets, from the urban planning perspective, may have locational attributes. An investment changes the attributes of the asset by bringing it into being or otherwise modifying it. Thus a decision to build a new road has spatial attributes by the virtue of spatial attributes of the road. Regulations are more akin to policies, which need not specify ahead of time particular locations to which they apply. However, regulations may have a spatial scope of applicability, typically a subset of the spatial scope of the jurisdiction of the actor who is regulating. The policy of choosing only two connecting streets between the collectors (Fig. 3(c)), for example, eliminates the choice of building three or one sub-collectors. Further, the policy is also about maintaining connectivity with other roads. It fixes the location of the intersections as and when new roads get built. A regulation specifying the minimum size of the lots in a particular zoning classification implicitly restricts how close the sub-collector streets can be, and thereby eliminates certain alternative locations from consideration.

Persistent debates about representations of geography, for example, object-field, crisp-vague dichotomies are particularly relevant to planning [10–12, e.g.]. In particular, the representation of inherent uncertainty about the location of the right of way of the sub-collectors can be represented as a probability field that exists between the two major collectors and gets modified by various events. The intended alignment of the Olympian drive is uncertain until the right of way is acquired by the city. However, geography is not central to the ontology for urban planning purposes. Location is but one attribute of urban development objects. The location of the effects of the decisions can provide a clue to which decisions might be related. However, other aspects of decisions, for example, jurisdiction of actor, which may not have a locational attribute, can also provide indicators to supplement understanding of how the decisions are related. In some cases, abstractions of locations are useful in determining the relationships between decisions. In Fig. 3(c) the notion of connectivity of a network of roads help narrow the alternatives where sub-collector should be built.

3.2 Alternatives

Alternatives are mutually exclusive actions. The exclusivity arises either because of capability constraints of actors or locational constraints of situating the action in a spatio-temporal setting. Keeping track of alternatives as they are modified, discarded, and used in a decision making process is useful because, as illustrated above, reporting intentions requires information about alternatives. In many urban development processes, these alternatives must be considered in a 'public' planning process. Alternatives can be of different types because:

- Multiple entities cannot occur simultaneously.
- 'Same' entity cannot happen in multiple instances.
- Multiple things may not occur in the same place.
- Same purpose can be achieved by different actions.

The possible locations of the interchange in the earlier example are alternatives. One interchange at one location can be built, but not all three because they are intended to serve the same purpose and they would create traffic conflicts if built close together. But in considering where to build the new interchange, it should be noted that there are three alternatives, which were considered at the time of the plan to dominate all other choices of locations, while no one of these three alternatives dominated the other two. In the future, as additional decision situations occur, this set of available alternatives may change, be reduced or expanded.

It is not useful to think of these three alternatives as separate decisions, to build or not to build each one, because they share an intention. The decisions are alternatives *with respect to* each other. I can either build the interchange at

A or I can build it at B or at neither place. That is, if I decide to build an interchange at A, I automatically also decided that the interchange at B is not going to be built in this particular context. A particular alternative thus has to include relationships with other alternatives.

If a plan specifies an alternative action in recognition of other intentions, then the planning process has recognised that other plan and represents its intentions about actions in its knowledge base. For example, a transportation plan might specify building extra lanes on an interstate highway whereas a plan by the local business organisation, in a directly contradictory approach, specifies that the rail network should be strengthened instead of building the lanes. Implicitly these are alternative uses for the same budget capacity toward the same intent for accessibility. When one plan recognises that the other includes an alternative action set, then a locational query could recognise the semantic relationship of alternatives in the two plans.

A plan might also specify multiple possible locations for the same road. While the recognition of 'sameness' of two proposals is not a trivial endeavour, it is possible that the plans might recognise these actions as the 'same' either in their intent or in their effect [13, 14, e.g.]. Thus intentions or effects can be used to identify the existence of alternatives.

In most cases, however, plans are circumspect about alternatives. To recognise that two actions are alternatives, expert knowledge about the situation is usually required. Such knowledge might involve, for example, recognition of budgetary constraints, which may preclude pursuing one kind of action when pursuing another. There may not be sufficient budget or borrowing capacity to build a new fire station and a new highway interchange, which become alternatives with respect to budget even though they are not alternatives with respect to intended purpose. The knowledge about 'priors', which are necessary and cannot be pursued simultaneously, might be involved to recognise the actions as alternatives. It may not be possible to build a new subdivision, for example, until the sewer services are extended. We can attempt to recognise the alternatives from the issues of location in a geographic context, location in a temporal context, and responsibilities actors and capabilities of actors, including their jurisdictions and budgets.

In all of the above examples in Urbana, the decisions have winnowed out various alternatives, which affect the implied commitment to a particular alternative. In the case of Fig. 3(c), the alternatives are effectively uncountable, subject only to policy restrictions on the distances between two parallel roads. Figure 3(b) considers three alternatives for future decisions, and Fig. 3(a) has specified a particular alternative as a decision. These differences can be viewed as differences in the types of commitment, by the deciding actor, to a particular set of alternatives [15]. A decision is an expression of a level of commitment to action. If we monitor whether or not an action is taken after the decision is taken, we can track the commitment of the particular actor to decisions and, conceptually at least, derive empirical estimates of commitment. More likely, we will use subjective estimates of commitment analogous to subjective probabilities.

3.3 Policies & Regulations

Information about assets, actors that hold rights over these assets, and regulations or transactions that change those rights is pertinent information for planning. An ideal system would track these changes of assets and changes of rights over these assets to arrive at "plan ready information" [16]. If we postulate that agents are planning continuously by amending old plans, updating them or discarding them in light of new information, relevant information about decisions needs to stay current.

Regulations are 'If-then' rules [17]. The 'Antecedent' describes the conditions when the particular regulation will hold and the 'Consequent' describes the rights through permission or by denial. Even when the regulations are performance based, the consequent can be used to describe the rights. Storm water runoff, for example, is often regulated to preclude any kind of development that alters the runoff characteristics of the site, thus circumscribing certain rights. Such regulations specify the attributes of effects of actions, thus giving wider latitude than regulations that specify a set of permitted actions in their consequent. In order to determine if a particular action is permitted or prohibited by the regulatory regime, it is then necessary to check not only the attributes of the action satisfying the regulation, but also the effects.

Rights have spatio-temporal dimensions. For example, sale of a property is an action that changes the rights of a current rights holder. The State can enact regulations about how this sale of property can be executed and what procedures should be complied with so that the State will guarantee this transaction, all without specifying when exactly the sale would occur. Hence, representing the time of the sale of the property is not sufficient to describe rights. We should be able to represent these events—sale, regulation, leasing, renting, and taking by the government—which routinely alter the set of rights and transfer these rights to other parties.

Policies are different from regulations because regulations are codified by statutory provisions and policies are merely a decision rule that gets applied repeatedly. A policy is chosen in anticipation of occurrence of repeated decision situations of similar kind. A policy could be announced for the sake of maintaining credibility, so that similar situations would be responded to in similar ways. However, policies and regulations share the same structural relationships between the antecedent and the consequent and thus could be modelled in similar fashion.

4 Relationships among Actions and among Decisions

The meaning of a decision changes when an action specified by the decision is carried out or when another explicit decision renders the earlier decision ineffective, perhaps by reducing its commitment to zero. We should keep track of these types of interactions for decisions and actions.

The difficulty of specifying the identity of objects, is also evident in specifying the identity of the decisions. For example, if a city annexes adjacent property into

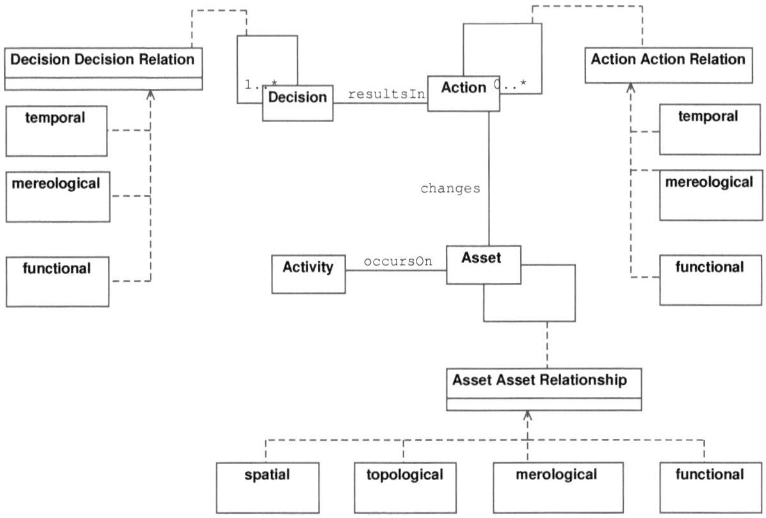

Fig. 4. Relationships among Decisions, Actions and Assets

its own jurisdiction, has the identity of the city changed? Similarly, a decision to build an interchange at either of the three locations is modified at a later point, by another decision that actually chooses the location. It is sometimes important to keep track of the sequences of decisions that resulted in the action to discover patterns of intents and effects.

Decisions do not typically happen in isolation but are linked to one another. The linkages can be temporal such as two decisions that have to be taken simultaneously. Or they can have spatial relationships such as intended investments that must be spatially adjacent. They can also be contingent or interdependent as shown in Fig. 4.

Almost all temporal relationships that are between actions happen between decisions [18]. However, the translation is not unique. For example, an action of relocating Olympian Drive is *followed by* the action of giving up the right of way of the current Olympian Drive (a finish-start relation among actions). The relationship between the two decisions, however, is *simultaneous*. The decision to build one interchange in East Urbana is followed by another decision about the location of the interchange. The result, however, is a single action.

Two decisions may be made by different actors but they may share a temporal relationship. An obvious example is a sequential play in a game theoretic sense between two actors. Consider the relationship between a decision of the Federal government funding of the construction of levees and a decision of a speculative developer to invest in the flood prone area. The speculative investment may occur prior to the building of the levees action or even before the decision to build the levees. A small homeowner, who is risk averse, may require more assurance about the flood protection of the area and thus may wait to rebuild a home until either there is a credible commitment to building the levees or even until after the levees get built. These relationships, which depend crucially on the notion of decisions as levels of commitment, have to be identified ahead of time and formulated as policies or strategies with which to monitor other's decisions to trigger one's own decision situations. The policies and strategies form the plans of the particular actor. Thus, decisions may be prior (and thus interdependent) on other actions or other decisions. Actions, by the same token, are interdependent on other decisions and actions.

If decisions are perfectly separable from each other, then the decision making process is simpler. However, most urban land development decisions ought to consider the effects of decisions of other actors and effects one's own action.[1] A zoning change near a proposed interchange is not very effective unless the interchange is built. Speculative investment in that parcel of land to develop it into a commercial strip, while purchasing the land when it is still zoned and used as agricultural land, necessarily depends on information about *what* decision about the interchange is likely to be taken and *when* . This information by its very nature is imperfect and subject to revision.

Typically, when a decision is taken, many implicit decisions are also taken. For example, to decide to build a new school would already imply commitment to, among other things, specify a location, provide infrastructure, staff it, and seek budgetary approval. In this sense all these decisions are encompassed in the decision to build a new school. However, it is unwise to assume that all such subsequent decisions are considered in complete detail and resolved in the current decision making process. The status of the decision—the alternatives that are chosen, timing, actors interested etc.—has a ripple effect on other decision situations that are yet to come, and sometimes that have already passed. In these situations, this ontology for urban planning, which takes into account the substantive knowledge about how plans, actions and actors work, will be useful.

The decision to build a new interchange at a particular location (Fig. 3(b)) is not a decision of one actor. The Federal government through its Department of Transportation, and the state's department of transportation must also decide to fund the project. The metropolitan planning organisation has to conduct a study about the traffic and other impacts of the project. The county and the city governments have to budget their shares of funding. As such, this decision is a decision-set by an ad hoc collection of actors. If any of those actors decides otherwise, the action is prevented from being taken. In particular, even after the decision is taken to build an interchange at the particular location, for example by the City of Urbana, the responsibility of carrying it through may lie with another actor, who is not involved in making the decision.

[1] A decision may be taken by a collection of actors agreeing to it, by various decision rules, including majority or unanimity. Such an actor would be an organisation or a collective. See [5].

5 Conclusion

A description of a continually developing ontology is available at `http://www.rehearsal.uiuc.edu/projects/pml/`. An ontology of urban development is necessary for building an Information System of Plans (ISoP), which should include substantive knowledge about how planning affects decision making and vice versa. An ISoP allows us to use multiple plans in decision making and modify plans continually to keep them relevant. This ontology enables sharing of information when authority and capabilities are distributed among disparate actors.

References

1. Keita, A., Laurini, R., Roussey, C., Zimmerman, M.: Towards an ontology for urban planning: The towntology project. In: CD-ROM Proceedings of the 24th UDMS Symposium, Chioggia (2004) 12.I.1
2. Hopkins, L.D.: Urban Development: The logic of making plans. Island Press, Washington, DC (2001)
3. Laurini, R.: Information Systems for Urban Planning. Taylor & Francis, New York, NY (2001)
4. Bratman, M.: Intentions, Plans, and Practical Reason. Harvard University Press, Cambridge, MA (1987)
5. Hopkins, L.D., Kaza, N., Pallathucheril, V.G.: Representing urban development plans and regulations as data: A planning data model. Environment & Planning B : Planning and Design **32**(4) (2005) 597–615
6. Rinner, C.: Argumaps for spatial planning. In Laurini, R., ed.: Proceedings of the First International Workshop on TeleGeoProcessing, Lyon, France (1999) 95–102
7. Knaap, G.J., Ding, C., Hopkins, L.D.: The effect of light rail announcements on price gradients. Journal of Planning Education and Research **21**(1) (2001) 32–39
8. Urbana: 2005 Comprehensive Plan. City of Urbana, Urbana, IL. (2005)
9. Couclelis, H.: Where has the future gone? Rethinking the role of integrated land-use models in spatial planning". Environment & Planning A **37**(8) (2005) 1353–1371
10. Couclelis, H.: People manipulate objects (but cultivate fields): Beyond the raster-vector debate in GIS. In Frank, A.U., Campari, I., Formentini, U., eds.: Theories and Methods of Spatio-Temporal Reasoning in Geographic Space, Berlin, Germany, Springer-Verlag (1992) 65–77
11. Campari, I.: Uncertain boundaries in urban space. In Burrough, P.A., Frank, A.U., eds.: Geographic Objects with Indeterminate Boundaries. Talyor & Francis, London, UK (1996) 57–69
12. Worboys, M.F., Duckham, M.: GIS: A Computing Perspective. 2nd. edn. CRC Press Inc., Boca Raton, FL, USA (2004)
13. Wiggins, D.: Sameness and Substance. Blackwell, Oxford, UK (1980)
14. Mark, D.M., Smith, B.: Ontology and geographic kinds. In: International Symposium on Spatial Data Handling, Vancouver, Canada (1998) 308–320
15. Levin, P.H.: Government and the Planning Process. George Allen & Unwin Ltd., London, UK (1976)
16. Carrera, F.: City Knowledge: An Infrastructure for Urban Maintenance, Management and Planning. PhD thesis, Massachusetts Institute of Technology (2004)

17. Kaza, N.: Towards a data model for urban planning: Ontological constructs for representing regulations and guidelines with geography. Master's thesis, University of Illinois at Urbana Champaign (2004)
18. Allen, J.F., Ferguson, G.: Actions and events in temporal logic. In Stock, O., ed.: Spatial and Temporal Resoning. Kluwer Academic, Boston, MA (1997) 205–243

An Ontology-based Model for Urban Planning Communication

Claudine Métral[1], Gilles Falquet[2], Mathieu Vonlanthen[2]

[1]Institut d'architecture (IAUG), University of Geneva
7, route de Drize, CH 1227 Carouge, Switzerland
claudine.metral@archi.unige.ch
[2]Centre universitaire d'informatique (CUI), University of Geneva
24, rue Général-Dufour, CH 1204 Geneva, Switzerland
{gilles.falquet, mathieu.vonlanthen}@cui.unige.ch[1]

Abstract. Urban planning projects are complex and involve multiple actors ranging from urban planners to inhabitants. These actors differ greatly in their background or their centres of interest. The main objective of our research is contributing to a better communication of urban planning projects between the various actors involved. With this intention, we defined an ontology-based model whose main characteristics are, on the one hand, the semantic integration in a knowledge base of the urban knowledge coming from various sources such as GIS databases, master plans, local plans or any other document and, on the other hand, the modelling of the centre of interest of an urban actor. This models can then be used to generate adapted user interfaces to present the project's data and knowledge according to each actor's background and interests.

Keywords: Urban planning, ontology, knowledge base, semantic integration, 3D city model

1 Introduction

Urban planning is a complex process involving many actors, such as urban planners, inhabitants, employees of urban technical departments, or politicians, and broad range of interests and demands. For instance, inhabitants are more and more implied in the way their city, and hence their way of life, is intended to change. But communication between the different actors is not easy. For example a plan, which is an obvious tool for an urban planner, may be hard to understand for the general public. Conversely, tri-dimensional (3D) representations are usually preferred by non-specialists, but are seldom used by urban planners. Moreover urban actors naturally use different terms and are interested by different types of information. For example a textual technical report with technical vocabulary is useful for an urban technical employee but not for a politician or an inhabitant who will probably misunderstand it. Finally, the actors

[1] This research has been funded by the swiss *Secrétariat d'Etat à l'Education et à la Recherche* as part of the COST C21 action.

are not necessarily interested in all aspects of a project and they may be interested in different aspects at different times. For instance, an actor may want to explore elements of a project that are related to water management, then he or she may be interested in data related to the safety on streets.

Thus, an efficient communication tool for urban planning projects must provide each actor with the information that is relevant for him or her and present this information in a way that is easily understood. This implies that the tool must take into account the user profiles and their centres of interest to present different views on the project.

As a contribution to improve the communication between the various actors involved in urban planning projects, we propose in this paper an ontology-based model that can serve as a basis to develop computerized tools for exploring and understanding urban projects. This model has two main components:

1. An integration component that is intended to represent in the same knowledge base information coming from different heterogeneous sources. This component is build around an ontology of urban planning process
2. An adaptable interface component whose aim is to provide each actor with a view of the urban project that corresponds to his or her profile and centre of interest. This component includes actor specific ontologies (viewpoint ontologies) and an ontology of themes

This paper is organized as follows: in the next section we briefly present tools that are currently used to represent data and knowledge in urban planning projects; in section 3 we introduce the notion of ontology and its applications in the urban planning domain; section 4 introduces the information integration part of our model and section 5 presents the user interface part of the model; section 6 gives our conclusion and perspectives for further work.

2 Urban Planning Knowledge and Tools

Working on an urban planning project involves working with heterogeneous and disseminated information obtained from various sources. These sources can be geographic information systems, master plans or local plans, legal texts, regulations, and, more recently, 3D city models.

2.1 Geographic Information Systems

A geographic information system (GIS) is essentially an information system that is intended to manage geographically-referenced information. It is usually comprised of a database system that stores geometric entities and can perform geometry or topology-based search operations. In the last decades, a lot of GIS containing spatial urban data have been created such as the *Système d'information du territoire genevois* (SITG) in Geneva. Such systems are very useful for example to obtain information on a parcel (owner, building) or to visualize information on a map (cycle paths, parking places reserved for handicapped persons, polluted sites). But GIS generally provide information about what exists, they cannot be considered as planning tools.

2.2 Master and local plans

Master plans are legal tools for the global planning of the territory. In Geneva the Cantonal Master Plan *(Plan directeur cantonal)* is the outcome of an extensive political and technical process to define in a consensual way the aspirations of the population concerning regional planning for the next fifteen years. Such a master plan is organised in different parts: texts, thematic maps (mobility, nature and landscape, etc.) and a synthesis map. The Cantonal Master Plan is available on the official web site of the State of Geneva. But, due to its density and its scale (the whole State of Geneva), it is difficult to retrieve precise information in it.

Local plans are legal tools for the planning of urban area under development or mutation. In Geneva they are subject to a public consultation. They are usually composed of texts and plans sometimes enriched with views and sections. If plans are usual tools for urban planners it is not true for the general public who often feels more at ease with 3D representations. Moreover local plans are not organized in the same way as the cantonal master plan, making difficult the crossing of information between master and local plans.

2.3 3D City Models

3D representations of urban data are named 3D city models. Different projects that model an existing city have been developed or are under development around the world. They are intended for a wide range of applications, such as planning and design, infrastructures and facility services, marketing or promotion [10].

3D city models can be built from existing GIS, which contain basically 2D information. For example by combining and extruding different SITG information layers, such as the digital terrain model (representing the ground without the vegetation or the buildings), the building footprints and the building heights, we obtain a 3D block model of Geneva. Strictly speaking such a model where the third dimension is expanded from 2D data using heights is a 2.5D model, but we refer to it as a 3D city model.

By adding a texture mapping from the orthophotos of the area we obtain a more realistic 3D model. More generally 3D city models differ by elements such as their degree of reality, i.e. the amount of geometric details that are represented within them, their data acquiring methods and their functionality, i.e. the degree of utility and analytical features that they allow [10].

Fig. 1. 3D city model extracted from the SITG

CityGML [7] is a unified model for the representation of 3D city models based on the standard GML3 of the Open Geospatial Consortium. Urban objects (relief, buildings, vegetation, water bodies, transportation facilities, city furniture) are represented in CityGML by features with geometric, topological and thematic properties. CityGML ensures spatial consistency between 3D models at different scales (five levels of detail are possible).

3D city models are useful for the visualization of full urban environments including built and natural structures or for the simulation of new urban projects with their environmental and visual impact. They are also a visual communication tool much more efficient than, for example, official plans. But what they represent is essentially of geometric nature when a lot of urban knowledge do not correspond to geometric entities (building period, parcel owner, building permit for example). Moreover 3D city models are not directly linked to the texts that are the main part of the master and local plans.

3 Proposal for an Ontology-based Model for Urban Planning Communication

An efficient communication of urban planning projects must, on the one hand, integrate the whole of the urban knowledge resulting from the various sources that are GIS databases, master plans, local plans, or any other document and, on the other hand, take in account the centres of interest of the different users. A centre of interest is in fact defined by a theme (mobility, environment, etc.) and the viewpoint that the user wants to have on this theme. Both the semantic integration of the urban knowledge and the specification of a user's centre of interest can't be based only on terms. The underlying semantics must be taken into account because some data and documents can be semantically related without containing the same terms. From where the idea to use ontologies.

3.1 Ontologies

In the field of Artificial Intelligence several definitions of the term "ontology" have been given. According to Gruber an ontology is an "explicit specification of a conceptualization" [6]. A slightly different definition is "a formal, explicit specification of a shared conceptualisation" [13]. A conceptualization is an abstract, simplified view of some domain that we wish to represent for some purpose, i.e. the objects, concepts and other entities that are assumed to exist in some area of interest and the relationships that hold among them. "Formal" means that some formal representation language has been used and so that the ontology is machine-readable and machine-processable. "Explicit" means that both the type of concepts used and the constraints on their use have been defined [2]. "Shared" refers to a common understanding of some domain that can be communicated across people and computers [13]. Three important areas where ontologies could be used have been reported: communication between people with different needs and viewpoints, interoperability between heterogeneous systems and systems engineering [15].

3.2 Urban Planning Ontologies

The urban planning field is concerned with ontologies. In Urban Civil Engineering, some preliminary experiments of the Towntology project have been made, producing ontologies such as an ontology for street planning and mobility [14]. Another project is related to buried urban infrastructure with as main focus the routing/alignment of an infrastructure network in such a way that minimizes its conflicts with other systems [9].

3.3 Ontology-based Communication of Urban Planning Projects

From our part, we use an ontology approach both for integrating the different data and documents related to urban planning projects and for enabling a user to define the best interface that fits his requirements and wishes.

4 Semantic Integration of the Urban Knowledge

One of the aims of our approach is to provide the user with an integrated view of an urban project. As we have seen above, the information about an urban project is represented in different media (databases, documents, 3D city models, etc.) and at different scales (from local plans to master plans). To integrate these information sources we propose to use a domain ontology of the urban planning process (OUPP) as a common conceptual reference and to connect the information sources to this ontology. We will thus obtain a knowledge base that represents the urban project.

This knowledge base is composed of:

- a conceptual layer (the OUPP) that describes all the concepts that appear in the documents and in the GIS database;

– a factual knowledge layer that links these concepts to the information sources.

There are two kinds of links between the information sources and the knowledge base:

– a *conceptual annotation link* connects an information element (a document or a part of a document) to a concept because this element mentions or is about this concept. For instance, a document about transportation would probably be connected to concepts such as “road”, “street”, “bus”;
– an *instance link* indicates that an information element is a particular instance of a concept. For example, a given local plan document can refer to parcels 1807 and 1809. In this case the link must contain some way to identify the instance (here the parcel number) and to find it in the information source (e.g. with a database query or an XML pointer to a document element, etc.).

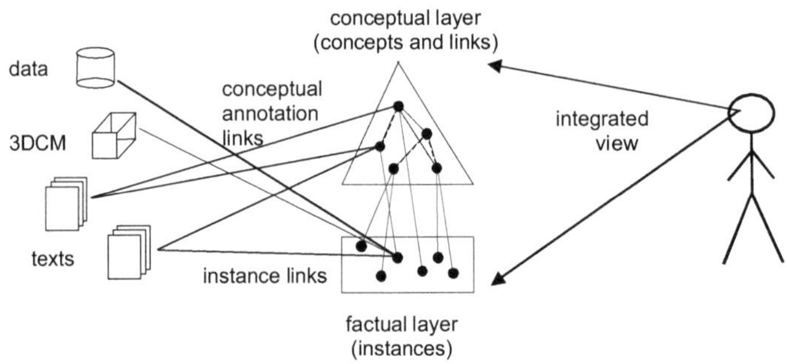

Fig. 2. Semantic integration of the urban knowledge

4.1 Ontology Construction

Research and practice in the field of ontologies showed that the construction of an ontology is a complex task requiring not only a great knowledge of the field to be described but also a control of the structuring of the concepts using formal languages. During the last years several approaches and tools have been developed to do these concept extractions automatically or semi-automatically. For instance [12] and [1] propose techniques to extract ontologies from relational database schemas, while [16] use text analysis technique to help in the construction of ontologies. At the same time, several languages have been developed to formalize ontologies, those being based primarily on predicate logic, on frames or on descriptive logic. The most recent works concern the language OWL which is a recommendation of the consortium W3C within the framework of the “semantic Web”. New tools and new methods for analysis of ontologies are under development [4].

We have built the first version of the OUPP with a simple graph editor then we have formalized it with OWL-DL. We have reused some parts of urban ontologies developed in the framework of the Towntology COST action (the *Ville* and *Transports* ontologies). We have established relationships to two of the main themes (“urban area”, “mobility” and “rural area”) of the Cantonal Master Plan of Geneva

(*Ville* has been related to "urban area" and *Transports* to "mobility"). Then we have added concepts from the master and local plans and then from the SITG whose concepts are less general. For the moment two kinds of semantic relations have been defined: the "isA" relation and the "isAssociatedWith" relation, which is symmetric and transitive.

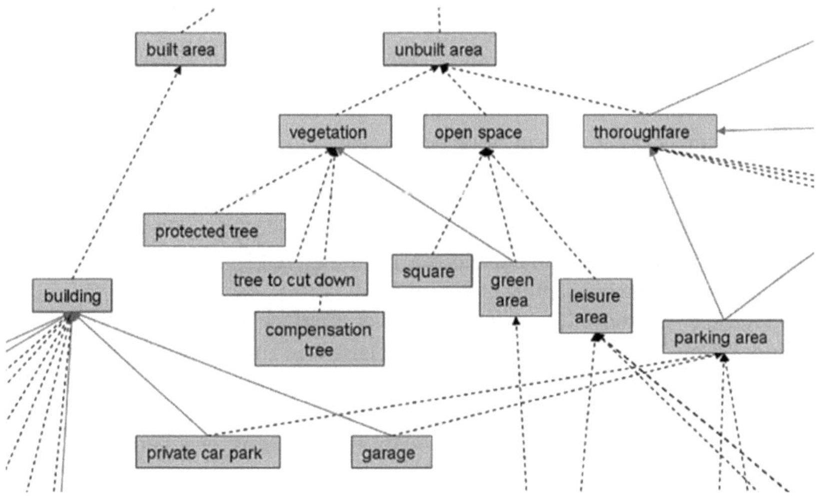

Fig. 3. Partial view of the OUPP (dotted lines represent *isA* links, solid lines represent *isAssociatedWith* links.)

4.2 Construction of the Factual Layer

Knowledge can be defined as organized information which can be used to produce new meanings and generate new data. According to Sowa [11] a knowledge base is an informal term for a collection of information that includes an ontology as one component. In our case, the construction of the knowledge base consists in linking the collected data (GIS data, natural language texts, and plans) to the ontology. The construction of the knowledge base corresponds to a semantic integration. For the GIS database [3] the concepts and relations of the ontology are directly inferred from the GIS database schema, So linking concepts and instances is immediate. For natural language texts, the linking problem is much harder. Concept extraction can be done using the terms associated with each concept. The main problem is as always in natural language processing (NLP) the polysemic terms. There exists a lot of different disambiguation techniques [8] but evaluating the different techniques and selecting the most appropriate is beyond the scope of this article. Instances extraction can be done readily when there exists object identifiers like local plan numbers or proprietary names. But when such information is not available and when the instance is defined in natural language as in “the house near the river and beyond the hill”, advanced NLP techniques are needed.

5 An Ontology-Based Adaptable Interface

5.1 Viewpoints

The actors involved in an urban planning process differ in many respects. In particular, they have different knowledge backgrounds and they use different vocabularies. To take account of this diversity we propose to represent it by different ontologies that correspond to the different types of actors. Each such ontology, called a viewpoint, represents the urban planning domain (or a part of it) as viewed by a given type of actor. It may of course differ from the OUPP in several aspects such as the terminology or the conceptual structure.

The idea is to use these viewpoints at the user interface level to:

- provide an interface that “speaks the user’s language”, i.e. all the interface elements such as menus, labels, etc. should use the user’s own vocabulary;
- provide navigation tools that the user immediately recognizes;
- display the information elements according to the norms, conventions or usage of this category of users.

To reach this goal, it is necessary to connect each viewpoint-ontology to the information sources. This is accomplished by establishing alignment links between each viewpoint and the reference ontology of urban planning process, as shown on the figure below. The alignment links interconnect the concepts of two different ontologies through an equivalence or a subconcept relationship.

The OUPP acts here as a “pivot language” among the different viewpoints. The benefit of this knowledge organization is that the links (factual layer) between the information sources and the OUPP are established only once and serve for all the viewpoints.

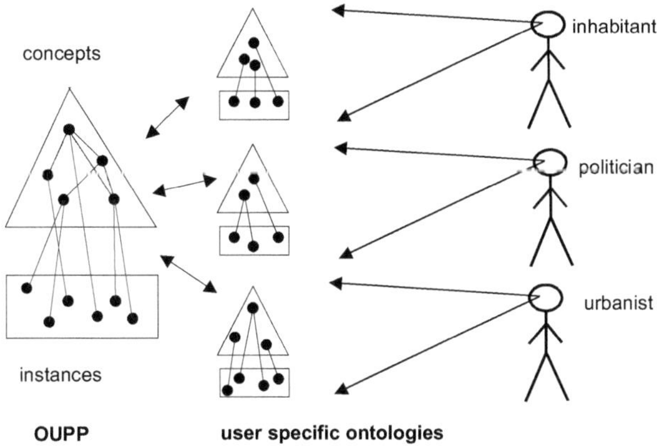

Fig. 4. Adaptation to user viewpoints

5.2 Themes

Within his or her own viewpoint, a user may be interested in different thematic aspects such as “transportation”, “safety”, “noise”, etc. These themes, or subdomains, correspond to sets of concepts and links in an ontology that do not necessarily form a connected subgraph of the ontology structure. Since these themes are, for the most part, common to all actors, we can represent them in a common ontology of themes. The concepts of this ontology can then serve to index the concepts and links of the OUPP. Hence, a theme *T* within the OUPP is made of all the concepts and links indexed by *T*.

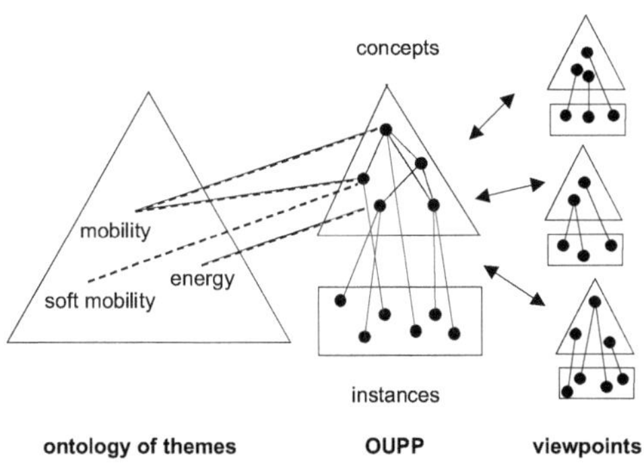

Fig. 5. Viewpoints and themes

At the interface level, the themes will serve as filters. By choosing a theme the user will restrict his or her view to the information elements that are relevant for this theme. Thus he or she will be able to concentrate on this theme without being disturbed by irrelevant information.

5.3 Adapted visualization

Ontologies have already been used for generating ontology-based interfaces. Among research works we can quote those realised at the University of Geneva [5] which generate hyperdocuments fitting the reading objectives or specific viewpoints of readers.

The ontology-based model we propose here is well suited to create adapted views of an urban planning project. This adaptation is based on the user profile, which is used to select a viewpoint-ontology *V*, and his or her current centre of interest, which corresponds to a theme *T* in the ontology of themes. Following the virtual document approach, the interface composition itself proceeds in two steps:

1. *V* and *T* determine the concepts that are relevant for the user and should appear on the interface. In addition, *V* provides the vocabulary that the interface must use for displaying information and for interface elements such as menus, lists of concepts, etc.
2. The conceptual annotation links and the instance links, together with the relations found in *V* and in OUPP give rise to visual and hypertext links between interface objects.

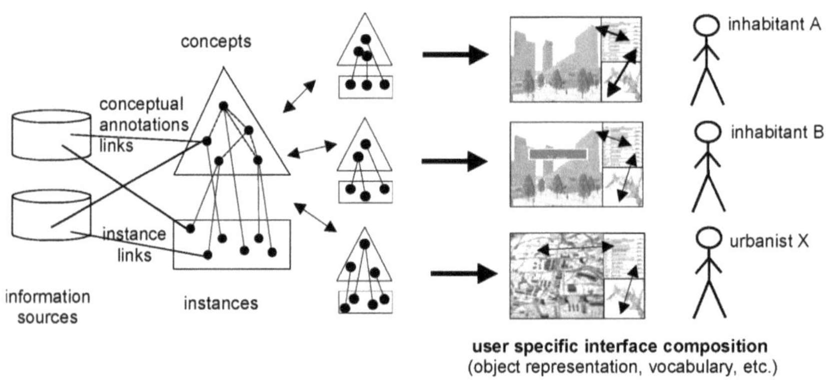

Fig. 6. Ontology-based user specific interfaces

The generation of links in the interface is carried out according to generation rules that correspond to different linking semantics. Typical rules are:

same instance: if the information elements *a* and *b* (belonging to different information sources) are both connected through an instance link to the same instance *x* of a concept *c*, then generate a link between the visual representations of *a* and *b*. This type of link is exemplified in figure 7 below, where the same object (building 107a) appears in a 3D view, on a plan and in a textual document.

Fig. 7. Ontology-based links in the interface. The solid rectangles visually interconnect representations of the same building in different views.

instance to concept: this is a type of link connecting the representation of an instance to its definition in the ontology V or to the definition of a related concept. For example a visualization related to "pedestrian path" can display not only such paths in the area of interest of the user but also the definition of "soft mobility" which is related to the concepts "pedestrian path" or "green path".

More complex rules may represent more sophisticated inferences involving the traversal of longer path in the knowledge base.

6 Conclusions and Future Work

In this paper we have presented a model that integrates in a knowledge base information and data from sources such as GIS databases, master plans, local plans or any document that seems to be relevant for the communication of an urban planning project. In addition to this semantic integration we specified an interface that fits a users's center of interest. We use an ontology-based approach both to the semantic integration and to the specification of the user interface.

The knowledge selected by the user by means of the ontology can be of various kinds: texts, plans, entities from the GIS database, etc. This knowledge is represented in our model as virtual documents that can take different forms ranging from hypertexts to 3D city models. However, there is still work to do, especially in finding and testing the best ways to represent non-geometric urban knowledge and linking documents with 3D city models. We also have to evaluate various interfaces with different urban actors and consequently to improve these interfaces according to their feedbacks.

References

1. Astrova, I. (2004). Reverse Engineering of Relational Databases to Ontologies, In: *Proceedings of the 1st European Semantic Web Symposium,* LNCS 3053, 327-341.
2. Benjamins, V. R., Fensel, D. & Gomez-Perez, A. (1998). Knowledge Management through Ontologies. In: *Second International Conference on Practical Aspects of Knowledge Management*, Basel, Switzerland.
3. Burrough, P. A. and McDonnell, R. A. (1998). *Principles of Geographical Information Systems*, Oxford University Press, Oxford.
4. Corcho, O., Fernández-López, M., Gómez-Pérez, A. (2003). Methodologies, Tools and Languages for Building Ontologies : Where is their Meeting Point ? In: *Data and Knowledge Engineering*, 46(1), 41-64.
5. Falquet, G., Mottaz, C.-L., Ziswiler, J.-C. (2004). Ontology Based Interfaces to Access a Library of Virtual Hyperbooks. In: *Rachel Heery, Liz Lyon (Eds.) Research and Advanced Technology for Digital Libraries. Proceeding of the 8th European Conference on Digital Libraries (ECDL 2004)*, Bath, UK, September 12-17, 2004. Lecture Notes in Computer Sciences (LNCS), vol. 3232, Springer, Berlin, Germany.
6. Gruber, T. R. (1993). A Translation Approach to Portable Ontology Specifications. In: *Knowledge Acquisition* 5(2), 199-220.
7. Kolbe, T. H., Gröger, G., Plümer, L. (2005). CityGML – Interoperable Access to 3D City Models. In: *Proceedings of the Int. Symposium on Geo-information for Disaster Management*, Delft, March 21-23.
8. Manning, C. D and Schütze, H. (1999). *Foundations of Statistical Natural Language Processing.* MIT Press, Cambridge, MA.
9. Osman, H. (2004). *A Knowledge-Enabled System for Coordinating the Design of Co-Located Urban Infrastructure.* Research Summary. Department of Civil Engineering, University of Toronto.
10. Shiode, N. (2001). 3D Urban Models: Recent Developments in the Digital Modelling of Urban Environments in Three-dimensions. In: *GeoJournal* 52 (3), 263-269.
11. Sowa, J. F. (2000). *Knowledge Representation: Logical, Philosophical, and Computational Foundations*, Brooks Cole Publishing Co., Pacific Grove, CA.
12. Stojanovic, L., Stojanovic, N., Volz, R. (2002). Migrating Data Intensive Web Sites into the Semantic Web, In: *Proceedings of the 17th ACM Symposium on Applied Computing*, 1100-1107.
13. Studer, R., Benjamins, V. R. & Fensel, D. (1998). Knowledge Engineering: Principles and Methods. In: *Data and Knowledge Engineering*, 25(1-2), 161-197.
14. Teller, J., Keita, A. K., Roussey, C., Laurini, R. (2005). Urban Ontologies for an Improved Communication in Urban Civil Engineering Projects. In: *Proceedings of the International Conference on Spatial Analysis and GEOmatics, Research & Developments, SAGEO, 2005* Avignon, France, June 20th-23rd.
15. Uschold, M. & Gruninger, M. (1996). Ontologies: Principles, Methods and Applications. In: *Knowledge Engineering Review*, 11(2), 93-155.
16. Velardi, P., Fabriani, P., Missikoff, M. (2001). Using Text Processing Techniques to Automatically Enrich a Domain Ontology. In: *Proceedings of ACM FOIS*, Ogunquit, Maine, USA, 270-284.

Towntology & hydrOntology: Relationship between Urban and Hydrographic Features in the Geographic Information Domain

Luis Manuel Vilches Blázquez[1], Miguel Ángel Bernabé Poveda[2], Mari Carmen Suárez-Figueroa[3], Asunción Gómez-Pérez[4], Antonio F. Rodríguez Pascual[5]

[1,5] Instituto Geográfico Nacional. General Ibáñez de Íbero, 3. 28003 Madrid. Spain. {lmvilches,afrodriguez}@fomento.es

[2] E.T.S.I. en Topografía, Geodesia y Cartografía. Universidad Politécnica de Madrid, Km 7.5 de la Autovía de Valencia. 28031 Madrid. Spain. {ma.bernabe@upm.es}

[3,4] Facultad de Informática. Universidad Politécnica de Madrid. Campus de Montegancedo, s/n. 28660 Boadilla del Monte. Madrid. Spain. {mcsuarez,asun}@fi.upm.es

Abstract. In this paper we describe the relationship between Urban Civil Engineering and other domains, specifically the hydrographic domain. The process of building hydrOntology and the portion of the model relating to urban features are described. This ontology emerges with the intent of settling as a framework in the GI domain, very closely interrelating to Towntology.

Keywords: GI (Geographic Information), Geographic Information System, Spatial Data Infrastructure, Ontological Framework, METHONTOLOGY

1 Introduction

Hydrography and related phenomena represent an essential part of reality in our cities as a consequence of the water supply needs they all have. This aspect is going to characterize some aspects of city planning owing to the presence of water infrastructures and to the addition of certain hydrographic phenomena in urban landscapes. This fact reflects the analogy of cities and other knowledge domains that, in view of their close relationship, are not irrelevant to the development of ontologies in the domain of Urban Civil Engineering. For that, a close collaboration between different scientific fields and disciplines is required, including civil engineering, urban design and planning and spatial information techniques [16].

These circumstances lead into an enhanced knowledge, since the use and development of ontologies are aroused in any domain Urban Civil Engineering projects are related to. This interrelation between different domains should contribute to enhancing access to GI.

Nowadays, in our society, the demand of GI is becoming a foremost need. Due to the poor, not well organized structure of GI as provided by the cartographic agencies, we come across many problems in the successful search and retrieval of data. These

L.M. Vilches Blázquez et al.: *Towntology & hydrOntology: Relationship between Urban and Hydrographic Features in the Geographic Information Domain,* Studies in Computational Intelligence (SCI) **61**, 73–84 (2007)
www.springerlink.com

problems mainly arise because each community producer is typically focused on specific needs [13]. That means that a harmonisation between the different agencies has not been achieved.

The development of Ontological Engineering is a key matter in the solution of current problems related to GI access and in distributed search in different cartographic organizations. For that reason, the definition of an ontological framework in the achievement of an easy accessibility and common structure of data becomes necessary. That means to provide a certain structure of names, codes, attributes and other associated represented characteristics being responsible for defining the real world. Thus, in order to give an answer to Society, these interrelated ontological frameworks (*hydrOntology and Towntology*) will hopefully improve the structure of the world of classical cartography, computer-assisted GIS (Geographic Information System) and SDI (Spatial Data Infrastructure).

With regard to *hydrOntology*, its purpose is to serve as a harmonization framework among the Spanish cartographic producers, trying to disseminate it internationally, making it available to GI producers. With this ontology we intend to provide the necessary steps to obtain a better organization and management of the hydrological features, which are spread over into the different projects, documents and directives in this field. To this information we should add a great number of catalogues, data dictionaries and so on due to the existence of different producers of GI. Another important characteristic is the different geometrical representation of the same domain (point, line, surface).

In section 2 of this paper, we describe the relationship between urban and hydrographic features. In section 3 we describe the problems encountered and characteristics of the integration process of the GI. Besides that, the semantic differences are commented in section 4. In section 5, the different ontological structure criteria are also commented, while, in section 6 the characteristic of the building up this *hydrOntology* through the use of METHONTOLOGY [1, 2] it is also given. Finally, in section 7, several conclusions and some future research lines are indicated.

2 Relationship between Urban and Hydrographic Features

Describing the richness of the urban environment in full detail represents a great challenge since this environment is very complex. It contains some natural occurrences like rivers that are features with natural boundaries. However, the urban environment is essentially made up of artificial objects. Even features such as rivers, when crossing urban environments, have their boundaries shaped by people and can be considered as artificial objects [19].

The change of hydrographic features into artificial objects is the result of the building of urban infrastructures for water supply, distribution and clean-up. Below three cases are shown where the close relationships between urban and hydrographic features are revealed.

1. The river feature has often been a key factor in the configuration of city maps. Because of this fact, urban infrastructures surrounding or being a part of this

feature are common. Actually we find retaining walls in river banks for canalization in a widespread fashion. Building of bridges as roads or passage ways between river banks is usual.

2. The water mains (piping) play a key role in the water supply, distribution and clean-up. Part of the mains are used for drinking water in the urban environment while another part of the pipes are utilized for residual waters that are channelled down to treatment plants for recycling and other uses.
3. Finally the sewer system and the rain water drains are most important for the urban environment owing to the fact that they take care of the removal of water from rainfalls or riverbed floods. Their efficient operation diminishes the effects of previous meteorological mishaps.

The close relationship between the features of both domains and consequently, the linking between these ontologies (*Towntology* and *hydrOntology*) will facilitate reaching the Towntology Project's aims. These are [17]:

- To identify terms and concepts used in different urban activities.
- To organize urban knowledge.
- To facilitate communication between various urban actors manipulating the same object types when achieving different goals.
- To gather urban data provided by heterogeneous sources.

In short, from the viewpoint of applicability, as a result of links between domains, the need to relate them becomes greater. This is due to the fact that, as a consequence of their interrelation, management of one of the utility networks can be set up. This is one of the subjects of INSPIRE [20], i.e. information referred to water supply and drainage networks (sewers, gutters, drainpipes, etc) could be controlled. Another interesting aspect coming up from the relationship between these two ontologies is the prevention of certain natural hazards affecting urban environments. Floods would thereby better managed and monitored due to the possibility of implementing applied hydrology models (estimation of maximum flows in the hydrographic network by means of empirical models) and through drainage models in cities in the face of unusually heavy rainfalls (statistical models).

3 Integration of Geographic Information

The basic unit of GI within most models is the 'feature', where by feature we mean an abstraction of a real world phenomenon, a geographic feature being a feature associated with a location on the earth [10]. Features can include representations of a wide range of phenomena that can be located in time and space such as buildings, towns and villages or a geometric network, geo-referenced image, pixel, or thematic layer. This means that traditionally a feature encapsulates all that a given domain considers about a single geographic phenomenon in one entity [9].

Features can be considered at two levels: feature instances and feature types. Feature instances are the individual discrete representations of geographic phenomena in a database with geographic and temporal dimensions. The instances may then be grouped into classes with common characteristics to form feature types. However, in Open Geospatial Consortium terms features are not fixed in their class but have application-oriented views that are classed [10] i.e. depending on the domain classification, a feature instance may be classified one way or another. Therefore, it is apparent that features are not the atomic units of GI as the phenomena they represent, encapsulating different human concepts resulting in multiple types [9]. This is the case of the hydrological domain, since there are different cartographic producers with various degrees of quality and structuring of information. That means a coexistence of a great variety of sources with different information and structure without a general harmonization framework.

In addition a scale factor should also be included which acts as a filter in the cartographic representation such as catalogues and dictionaries in the hydrological domain. For this reason, we have to consider information at several scales (local, regional and national) in the *hydrOntology*, though we are aware of the fact that in-depth work in the hydrographic features of cities should be carried out, owing to the change in geometric and semantic resolutions brought about by the scale difference between both domains. Moreover, some problems related with language ambiguity should be added, such as polysemy, synonymy, hyperonymy and homonymy present in many concepts in this domain.

An added drawback in the creation of *hydrOntology* has been the scarce semantic information present in many information sources consulted (EuroGlobalMap, EuroRegionalMap, DGIWG group FACC codes (*Feature Attribute Code Catalogue*), Numerical Cartographic Database to scale 1:200.000 and 1:25.000 of Instituto Geográfico Nacional of Spain (IGN-E), feature catalogues of Spanish cartographic producers, Geographic Gazetteer (IGN-E), etc. This information is of fundamental importance to distinguish and compare features in any knowledge domain.

Consideration of these facts has left a trace in the modelling process of this ontology framework of hydrographic features by trying to solve recurrent problems and contribute to shared knowledge.

4 Semantic Differences

The existing semantic differences in some domains are numerous, and this is so in the hydrographic domain, where several meanings and concepts are encountered. A repetitive example in this knowledge domain is the river definition. The Water Framework Directive (WFD) defines a river feature as "*a body of inland water flowing for the most part on the surface of the land but which may flow underground for part of its course*" [11], while the Ordnance Survey defines it as "*water flowing in a definite channel towards the sea, a lake or into another river*" [12]. On the other hand, the IGN-E considered the river a "*natural freshwater stream*". Nowadays, the IGN-E has decided to adopt the WFD proposal because it is a continuous phenomenon, although it would lack a cartographic representation when the flowing occurs underground.

Due to the diversity in semantic concepts within the domain, the definition of the characteristics and the context has been restricted, adapting it to the topographic data base, as the Numerical Cartographic Database of IGN-E. Every definition will take the cartographic representation into account through map, GIS or SDI, no matter what the intrinsic reality of these phenomena is.

Furthermore, in the *hydrOntology* development we have taken into account some concepts about feature capture which depend exclusively on different geographic regions, since they are concepts related to their importance in both Geography and Cartography. Among these features appear "ibón", "lavajo", "chortal", "bodón" and "lucio". These concepts are designated by their local name and they are synonymous to the feature "Charca"[1] , i.e. a small lake of shallow water. Later, we will analyse other international GI catalogues and dictionaries, adding further concepts of this kind to enrich this ontology.

Finally, due to the mapping purpose of this ontology to other knowledge bases (Thesaurus of UNESCO[2], Alexandria Digital Library[3], Thesaurus GEMET[4], Getty Thesaurus of Geographic Names[5], etc.) several features are considered which will be used to relate to other domains, such as the legal framework (international law). Concepts like "territorial waters", "contiguous zone", "high seas", etc., or the geological domain (hydrogeology) "underground currents", "aquifers", etc. may be considered as an example.

5 Criteria for *hydrOntology* Structuring

Taking into account the difficulties related to ontological framework standardization as mentioned above, we propose *hydrOntology* as a concurrent model to solve the structuring and harmonization problems for the GI community.

The organization present in this ontology about hydrographic features is governed mainly by four criteria:

1. The European Directive to set up *a communitarian frame of performance in the scope of the water policy (WFD)* [11]. Precisely, in article 2, a list of hydrographic phenomena definitions is given which may be considered as an implicit classification. That contributes to the modelling of more abstract features that make up the *hydrOntology* taxonomy. The definitions of hydrographic phenomena gathered in this article are proposed by the European Parliament and the European Union Council which makes such proposals mandatory in any taxonomy within this domain.
2. On the other hand, as a consequence of the aim of implementation of this ontology in the SIGNA-E and in the IDEE, we are taking into consideration the classification worked out by the SDIGER Project[6] [18], [an SDI created to support the access to GI resources concerned with the WFD within an inter-administration and cross-border scenario that involves two countries, France and Spain as well as the two main river basin districts on both sides of the border, the Adour-Garonne

[1] The above mentioned terms are Spanish local names.
[2] http://www2.ulcc.ac.uk/unesco/
[3] http://www.alexandria.ucsb.edu/
[4] http://www.eioneteuropa.eu/GEMET
[5] http://www.getty.edu/research/conducting_research/vocabularies/tgn/
[6] http://www.idee.es/sdiger/

basin district, managed by the Water Agency for the Adour-Garonne River Basins[7] (*L'Agence de l'Eau Adour-Garonne*) and the Ebro river basin district, managed by the Ebro River Basin Authority[8] (*Confederación Hidrográfica del Ebro*)]. That project was chosen by Eurostat[9] as a pilot project of the applicability of INSPIRE. In addition to those documents [14], the development of *hydrOntology*'s modelling such as UML models [15] from the above mentioned SDIGER project have a strategic importance. In the phase of analysis those models were adopted and several changes were included reaching a consensus with the Working Group of the University of Zaragoza. Those changes upgrade the proposed models.
3. Being aware of the importance of the establishment of a taxonomical order, several semantic criteria have been added. Thus the hydrographic feature classification is in accordance with the meaning of each feature.
4. Finally, an important matter should be added to those criteria, namely the presence of the inheritance of different sources in the modelling of this ontology, on the one hand to facilitate the possible information mapping and on the other to be consequent with the hierarchy of the features carried out by the expert in the domain.

6 Characteristics of *hydrOntology* Development Process

The development of *hydrOntology* has been based on METHONTOLOGY which was developed within the Ontological Engineering Group (OEG)[10] at Universidad Politécnica de Madrid. This methodology enables the development of ontologies at the knowledge level, and has its roots in the main activities identified by the IEEE software development process [3] and in other knowledge engineering methodologies [4].

This methodology (*METHONTOLOGY*) has been used by different groups to build ontologies in different knowledge domains, such as Chemistry, Science, Knowledge Management, e-Commerce, etc. A detailed description of the methodology of this ontology building can be found in [2].

---

[7] http://www.eau-adour-garonne.fr
[8] http://www.chebro.es/
[9]http://epp.eurostat.cec.eu.int/portal/page?_pageid=1090,30070682,1090_33076576&_dad=portal&_schema=PORTAL
[10] http://parla.dia.fi.upm.es/oeg/

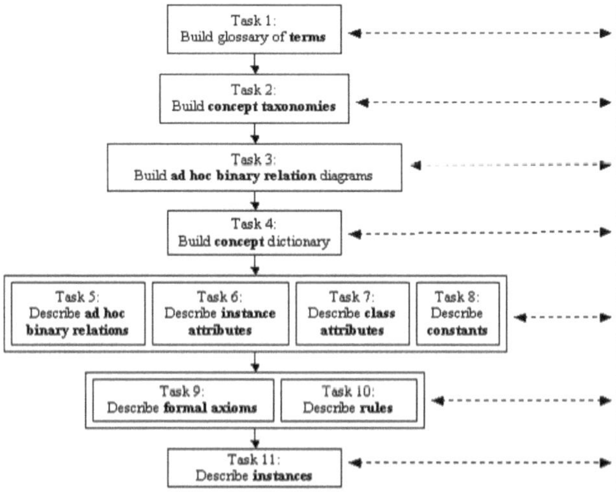

Fig. 1. Tasks of the conceptualization activity according to METHONTOLOGY[2]

In order to ensure the *hydrOntology* consistency and completeness, several steps have been followed. Figure 1 shows the ontology building tasks suggested in the METHONTOLOGY framework [5].

As seen in Figure 1, a glossary of terms was built, as a result of the study of several feature catalogues and data dictionaries (Numerical Cartographic Database of the IGN-E, catalogues and data dictionaries from other cartographic agencies, WordNet, etc.), thesauri (UNESCO, GEMET, Getty TGN, etc.), the project SDIGER, different classification systems and taxonomies (Alexandria DL, Dewey, etc.), etc., trying to cover the greatest amount of IG sources, in order to build a complete ontological frame. This glossary contains more than 100 relevant concepts related to hydrology as river, reservoir, lake, channel, pipe, water tank, siphon, etc.

In a first approach, a taxonomy of concepts was built. METHONTOLOGY suggests using the four taxonomic relations defined in the Frame Ontology [6] and the OKBC Ontology [7]: Subclass-Of, Disjoint-Decomposition, Exhaustive-Decomposition and Partition.

A concept C1 is a *Subclass-Of* another concept C2 if and only if every instance of C1 is also an instance of C2. [5].

A *Disjoint-Decomposition* of a concept C is a set of subclasses of C that do not have common instances and do not cover C, that is, there can be instances of the concept C that are not instances of any of the concepts in the decomposition [5]. An example of this type of relationship is shown in Figure 2.

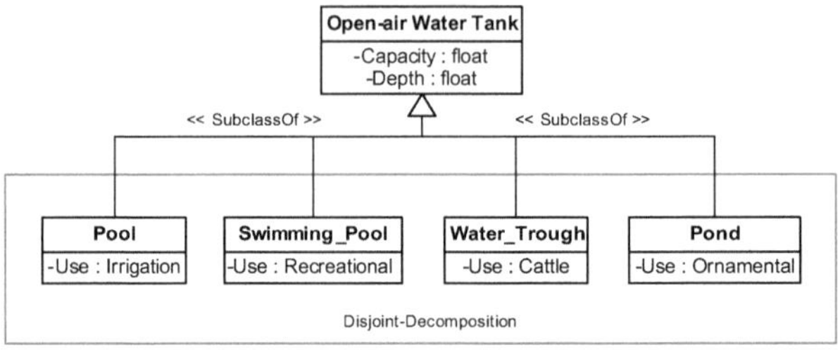

Fig. 2. Example of Disjoint-Decomposition included in *hydrOntology*

An *Exhaustive-Decomposition* of a concept C is a set of subclasses of C that cover C and may have common instances and subclasses, that is, there cannot be instances of the concept C that are not instances of at least one of the concepts in the decom-decomposition [5]. Figure 3 shows an example of this type of relationship.

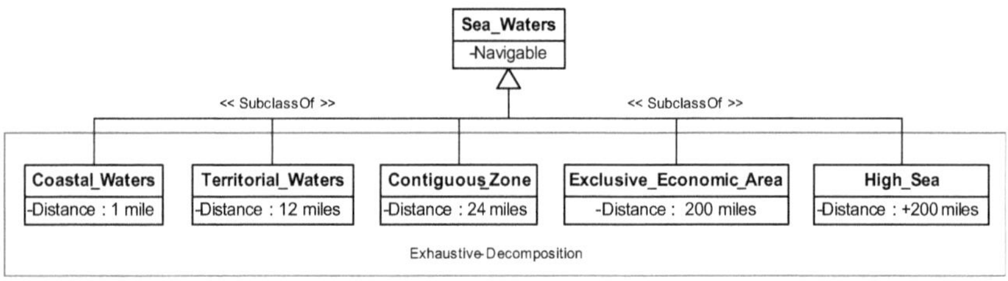

Fig. 3. Example of Exhaustive-Decomposition included in *hydrOntology*

A *Partition* of a concept C is a set of subclasses of C that do not share common instances and that cover C, that is, there are not instances of C that are not instances of one of the concepts in the partition [5]. An example of a partition is shown in Figure 4.

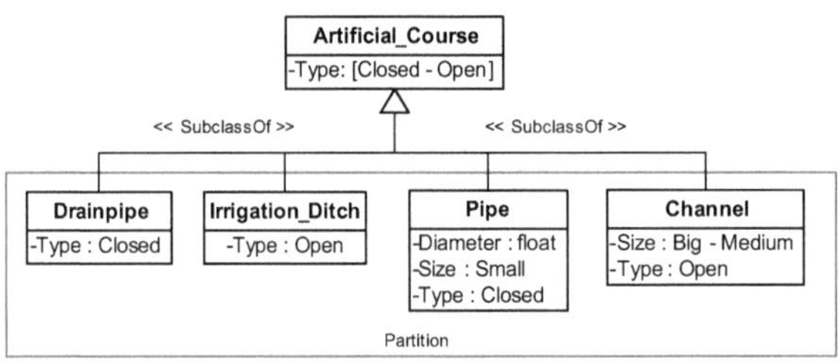

Fig. 4. Example of Partition included in *hydrOntology*

Once the different taxonomic relationships had been established and due to the absence of semantic information in many of the sources of information considered, a conceptual hydrographic dictionary was constructed. That implies endorsing GI semantics. Among different sources considered in the building of this dictionary we should mention WordNet[11], Encyclopaedia Britannica[12], Diccionario de la Real Academia Española de la Lengua[13], Wikipedia[14] and several geographical dictionaries. After carrying out those steps, we went through the taxonomy to make sure it did not contain any errors [8]. Moreover, a dictionary was drawn up and used to ensure that the taxonomic organization was semantically consequent.

Once the taxonomy of concepts was correctly structured, an ad-hoc relationship between different ontology concepts was established. The type of relationship and other components explicitly contribute to enrich the *hydrOntology*. An example of an ad hoc relation is shown in Figure 5.

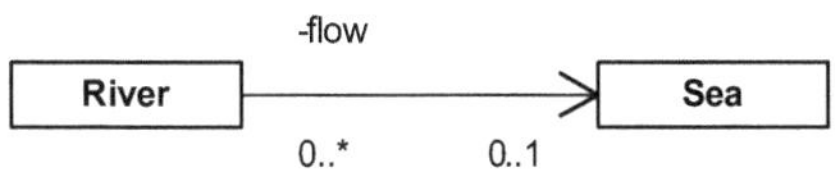

Fig. 5. Example of ad hoc relationship included in *hydrOntology*

Next step in the development of this ontology was the attribute specification for every concept. This is a difficult, subjective task due to the ambiguity and similarity of many real world phenomena. Then a differentiation between instance attributes and those belonging to the classes was applied. The instance attributes are those attributes whose value(s) may be different for each instance of the concept [2]. On the contrary, the class attributes describe concepts and take their value in the class where they are defined [2]. In Figure 2 and Figure 3, a clear example of this type of attributes is shown. On the one hand, in Figure 2 the instance attributes are shown by means of the information related to a specific value for each "`distance`" in the different subclasses and on the other hand the class attribute "`navigable`" as a class generic attribute. In Figure 3, the instance attributes are shown by means of the information related to a specific type for each "`use`", while the class attributes appear in the Open-Air Water Tank Class ("`Capacity`" and "`Depth`").

After having carried out the different steps, in view of the obvious implications and alterations involving urban and hydrographic features, the need to relate *hydrOntology* and *Towntology* was considered. The relationship between these ontological frameworks facilitates communication between various urban actors, organization and management of knowledge are improved [17] and a way toward a cooperative system is provided, capable of looking at knowledge in a scalar way, with added benefit for the users. Figure 6 shows an example of the relationship between the different urban and hydrographic features.

[11] http://wordnet.princeton.edu/

[12] http://www.britannica.com/

[13] http://www.rae.es

[14] http://www.wikipedia.org/

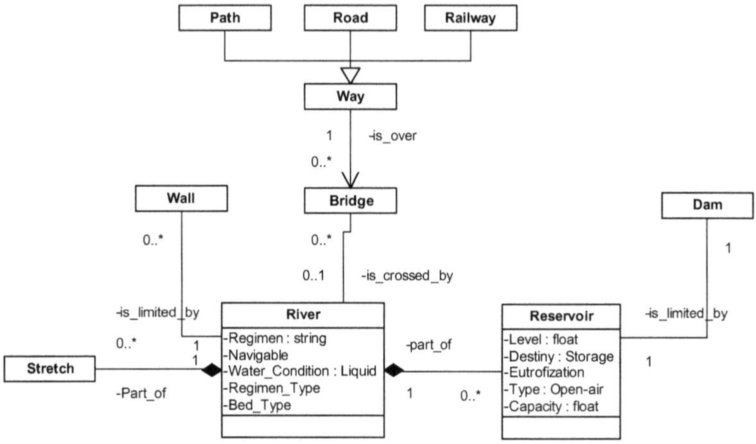

Fig. 6. Example of relationship between urban and hydrographic features

Finally once the conceptual modelling process has been taken care of for this ontology, we will try to include instances of different concepts that are part of *hydrOntology* and *Towntology*. Those instances will be gathered from the different national cartographic producers.

7 Conclusions and Future Work

Reflections on the relationships between Urban Civil Engineering ontologies and other domains become necessary, since a much greater knowledge and applicability are thereby achieved. At the same time these relationships allow promotion and improvement in communication between different information systems as a result of a better structuring of information, a better establishment of relationships and the possibility of feature mapping.

In this paper, all the different problems which reflect the difficulty to access to the GI are considered. This indicates that further structuring of information is needed as the complexity and volume of data increases. In other words, it is also necessary to have an ontological framework.

Although this ontology is in a stage of development, it constitutes an important headway towards an optimal structuring of semantic information by the spatial data producer organizations.

hydrOntology is also an interesting project in the IGN-E because it improves the information classification and management, in favour of the optimization in the search and recovery of the GI supported by the IDEE and the SIGNA-E. However, it tries to establish it as a generic semantic frame for use of every producer organization. This contributes to the shaping of a common, shared knowledge in the GI domain. With this ontology we think it is possible to define, relate and regulate the features in a unique way once a consensus is reached. We will do away with today's heterogeneity.

The next phase of this work will improve *hydrOntology,* by means of its enrichment through the implementation of possible rules, axioms and constants. In addition, the instances through compiled information from diverse sources will be added. With these processes, we are trying to draw inferences and a greater knowledge of the domain. We will also establish mapping to different knowledge sources (Digital Alexandria Library, Thesaurus of UNESCO, Wikipedia, etc.) and with ontologies of related domains such as *Towntology*, thereby extending the information of the features contained in this ontology framework.

Finally, we will analyze a number of feature catalogues and dictionaries from different worldwide organizations. This will serve as a starting point in the *hydrOntology* adoption as an ontology framework in the GI world.

References

1. Fernández-López M, Gómez-Pérez A, Juristo N (1997) METHONTOLOGY: From Ontological Art Towards Ontological Engineering. Spring Symposium on Ontological Engineering of AAAI. Stanford University, California, pp 33–40.
2. Gómez-Pérez A, Fernández-López M, Corcho O (2003) Ontological Engineering: with examples from the areas of knowledge management, e-commerce and the Semantic Web, Springer-Verlag, New York.
3. IEEE (1996) IEEE Standard for Developing Software Life Cycle Processes. IEEE Computer Society. New York. IEEE Std 1074-1995.
4. Gómez-Pérez A, Juristo N, Montes C, Pazos J (1997) Ingeniería del Conocimiento: Diseño y Construcción de Sistemas Expertos. Ceuta, Madrid, Spain.
5. Corcho O, Fernández-López M, Gómez-Pérez A, López-Cima A (2005) Building legal ontologies with METHONTOLOGY and WebODE. Law and the Semantic Web. Legal Ontologies, Methodologies, Legal Information Retrieval, and Applications. Springer-Verlag.
6. Farquhar A, Fikes R, Rice J (1997) The Ontolingua Server: A Tool for Collaborative Ontology Construction. International Journal of Human Computer Studies 46(6):707–727.
7. Chaudhri VK, Farquhar A, Fikes R, Karp PD, Rice JP (1998) Open Knowledge Base Connectivity 2.0.3. Technical Report. http://www.ai.sri.com/~okbc/okbc-2-0-3.pdf
8. Gómez-Pérez A (2001) Evaluation of Ontologies. International Journal of Intelligent Systems 16(3):391–409.
9. Greenwood J, Hart G (2003) Sharing Feature Based Geographic Information - A Data Model Perspective. 7th Int'l Conference on GeoComputation. United Kingdom.
10. OPEN GEOSPATIAL CONSORTIUM, 2003, OpenGIS Reference Model, Version 0.1.2, Open Geospatial Consortium Inc. Wayland, MA, USA.
11. European Parliament and the Council of the European Union. (2000) Directive 2000/60/EC of the European Parliament and of the Council for establishing a framework for Community action in the field of water policy. Brussels, p 72.
12. Ordnance Survey (2001): Master Map real-world object catalogue. http://www.ordnancesurvey.co.uk/oswebsite/products/osmastermap/faqs/Docs/realWorldObjectCatalogue.pdf
13. Bermudez L, Piasecki M (2004) Role of Ontologies in Creating Hydrologic Metadata. International Conference on HydroScience and Engineering, Brisbane, Australia.

14. Institute Geographique National France International, Institut Géographique National France, Centro Nacional de Información Geográfica Spain and University of Zaragoza. (2005) Common Model Activity - Final report. SDIGER: A cross-border inter-administration SDI to support WFD information access for Adour-Garonne and Ebro River Basins. http:/www.idee.es/sdiger/public_docs/CommonModelActivity_FinalReport_v0.3.pdf
15. Institut Géographique National France International, Institut Géographique National France, Centro Nacional de Información Geográfica and University of Zaragoza. (2005). SDIGER UML models. http://www.idee.es/sdiger/public_docs/ThemWaterUMLModels.pdf
16. Teller J., Keita A. K., Roussey C., Laur ini R. (2005), "Urban Ontologies for an improved communication in urban civil engineering projects", Proceedings of the International Conference on Spatial Analysis and GEOmatics, Research & Developments, SAGEO 2005 Avignon, France, June, 20th-23rd.
17. Keita A., Laurini R., Roussey C., Zimmerman M. (2004), Towards an Ontology for Urban Planning: The Towntology Project. In CD-ROM Proceedings of the 24th UDMS Symposium, Chioggia, October 27-29, 2004, pp 12.I.1.
18. Latre M.A., Zarazaga-Soria F.J., Nogueras-Iso J., Béjar R., Muro-Medrano P.R. (2005) SDIGER: A cross-border inter-administration SDI to support WFD information access for Adour-Garonne and Ebro River Basins. Proceedings of the 11th EC GI & GIS Workshop, ESDI Setting the Framework. Alghero, Sardinia (Italy).
19. Fonseca F., Egenhofer M., Davis C., Borges K. (2000) Ontologies and Knowledge Sharing in Urban GIS. CEUS – Computer, Environment and Urban Systems 24 (3) pp 232-251.
20. Commission of the European Communities (CEC) (2004) Proposal for a Directive of the European Parliament and of the Council establishing an infrastructure for spatial information in the Community (INSPIRE). COM(2004) 516 final, 2004/0175 (COD).

Visualizing the Uncertainty of Urban Ontology Terms

Hyowon Ban[1] and Ola Ahlqvist[1]

[1]Department of Geography, The Ohio State University,
1049B Derby Hall, 154 N Oval Mall
Columbus, OH 43210, USA.
{ban.11, ahlqvist.1}@osu.edu

Abstract. The concept of exurban is an example of a term likely to find its way into urban development ontologies. Many such terms are uncertain since there is no consensus of the exact definition of e.g. the boundary of exurbanization. In this research we focus on visualizing the spatial implications of the uncertainty in two existing definitions of exurban boundaries using empirical GIS data in Delaware County, Ohio, U.S. We argue that exurban boundaries are not crisp, hence, a series of fuzzy-set theory membership functions help define the uncertainty of the empirical exurban boundaries.

Keywords: uncertainty, visualization, exurban, boundary definition

1 Introduction

In Urban Civil Engineering (UCE) studies, there have been efforts to produce a taxonomy of ontologies, analyze the role of ontologies as a tool to foster an improved communication between stakeholders by building multi-lingual UCE glossaries of explication [24]. The studies emphasize that any serious attempt to construct urban ontologies must accommodate the evolution of concepts among different actors. This is because different groups have different concepts about the urban environment according to the inherently sociotechnical character of ontologies. Ontology also plays an essential role in the construction of Geographic Information Systems (GIS), since it allows the establishment of correspondences and interrelations among the different domains of spatial entities and relations [22].

Ontology, the science of being [21], is a logical theory accounting for the intended meaning of a formal vocabulary, and it determines what can be represented and what can be said about a given domain [24]. Fonseca et al (2000) analyze the urban environment from the ontologists' point of view. Other prospects suggested in the literature is composition of pre-existing independently developed ontologies, for instance, through the use of a context algebra to compose diverse ontologies [28], or through proxy contexts [2]. There are however a number of issues that make the application of ontology to urban areas problematic. Because of the differences in understanding concepts that form an ontology it is important that these differences can be articulated in some way. Some concepts in for example an urban area type ontology, terms such as urban, sub-/exurban, and rural areas, are inherently vague.

H. Ban and O. Ahlqvist: *Visualizing the Uncertainty of Urban Ontology Terms,* Studies in Computational Intelligence (SCI) **61**, 85–94 (2007)
www.springerlink.com

Yeates (1993) suggests 5 stages in the transition from exurban to suburban: agricultural, early urban influence, small town growth/exurbanization, and urban. Because the urban environment does not cease to exist abruptly—i.e., bona fide [20]—at the municipal borders, it is essentially fiat [20];[9] and should be treated as continuous.

Recently exurban areas have received specific attention because of its fast growth. According to Theobald (2005) in 2000, there were 125,729 km^2 of urban and suburban (<0.68 ha per unit) residential housing nationwide (conterminous USA), and about seven times that (917,090 km^2) of exurban housing (0.68–16.18 ha per unit). Statistics like these depend on definitions of what constitutes an exurban area and there are a number of suggested definitions. As there are various names to call exurbanization, most of them come with a separate definition. Irwin and Bockstael (2004) argue that since there is little consensus on a definition, data and measurements of sprawl are highly dependent on the researchers. Moreover, ontology is not likely to provide a 'silver bullet' simply because spatially continuous phenomena have received very little attention in the field of ontology [13]. There has been some research on boundary issues in exurban studies but few of them mention the uncertainty of the boundaries in exurban areas. Theobald (2005) models exurban land-use changes with a Landscape Sprawl (LS) metric. Wolman et al (2005) argues for measuring sprawl using data on density, concentration, centrality, nuclearity, and proximity of areas. Wilson et al. (2003) develop a model that determines the geographic extent, patterns, and classes of urban growth over time using land cover data. However, the exurban boundaries are defined crisp in their work. Caruso (2005) addresses the issue of urban expansion by exploring the emergence and morphology of a periurban zone at the periphery of a city, where residents and agricultural activities mix.

Generally uncertainty consists of errors, vagueness, and ambiguity [7]. Errors can be represented with probability, the vagueness can be explained by fuzzy set theory, and the ambiguity contains discord and non-specificity as its innate characteristics [7];[14]. Among the types of uncertainty, we think of vagueness when there is no unique allocation of individual objects to a class, or no precise spatial extent of the objects [7]. We can also find ambiguity when more than one definition for a term exist, one object is clearly defined but is shown to be a member of different classes under differing classification schemes or interpretations [14].

The purpose of this paper is to compare the spatial implications of different ontological commitments as they are represented by different definitions of exurban areas. We also want to demonstrate the relevance of representing exurban areas as vague objects by comparing the traditional crisp representation with a vague, graded representation. To do so, we represent the different theoretical boundaries of exurban areas using crisp boundaries and fuzzy membership functions and visualize these empirical boundaries of exurban areas in maps of Delaware County, Ohio, USA using standard GIS techniques.

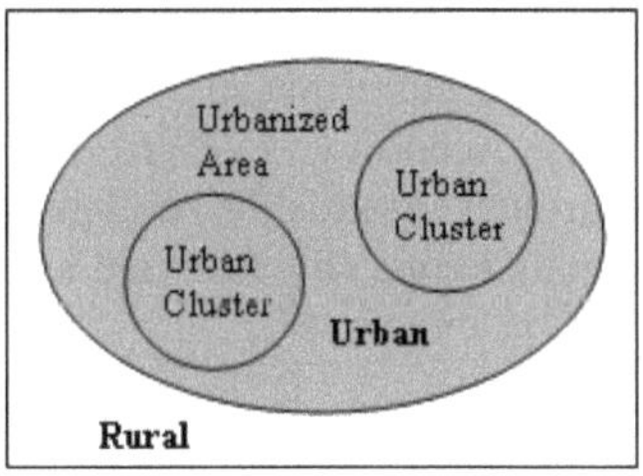

Fig. 1. Idealized spatial configuration of urban and rural area concepts

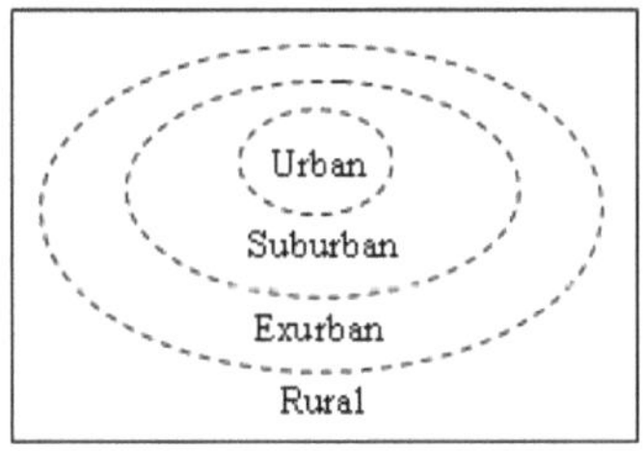

Fig. 2. Simple spatial distribution of urban, suburban, exurban, and rural areas

2 Theoretic background and concept of uncertainty in exurban boundaries

We start by introducing some ontological terms used to identify areas related to an urban environment.

2.1 The concept of urban, suburban, and rural zone

Urban means a zone—also called Urbanized Area—that has a population of at least 50,000 people and a population density of at least 1,000 people per square mile. Such zones are located within an urbanized area or an urban cluster (Fig. 1) [26]. Urban cluster means an area that has census block groups that is contiguous and densely distributed and its census block groups have at least 2,500 people but fewer than 50,000 people. Suburban county means a non-central county classified as metropolitan. Metropolitan counties outside this ring of suburban counties are considered exurban [16].

Rural zone means all areas outside the boundary of an urbanized area. Part of an administrative area—such as a census tract, or county outside metropolitan areas—can belong to urban area(s), and the other part of it can belong to rural area(s) simultaneously [26]. It is hard to say whether an administrative unit is 100% rural or 100% urban. Based on this, we get a simplified spatial distribution of urban, suburban, exurban, and rural zones as Fig. 2. The dotted boundaries between suburban-exurban and exurban-rural indicate that these areas don't have a clear boundary between them. Based on this, we might imagine a conceptual space consist of urban ontology terms. In that space, an axis starts from the term 'urban' and ends at 'rural'. On the axis, the term 'exurban' exist between them but more close to 'rural' with other similar terms such as 'sprawl' or 'periurban'.

Turning our focus on the exurban areas Daniels (1999) define this urban fringe as a region of middle ground between the wide-open rural lands that are beyond commuting distance to a metro area and the expanding suburban residential and commercial developments. Again, as indicated in Fig. 2, it is not easy to define the exact location of exurban areas. The exurban area is also called ‘periurban’ mainly as in the French-speaking literature, ‘deconcentration’, ‘decentralisation’, or ‘extended suburbanisation’, in Europe and North America [4]. According to Caruso (2005), periurbanization refers to the process of residential growth towards the rural periphery of a city. This process leads to the emergence of a spatial zone characterized by a mix of agricultural activities and commuting households [5] (cited in [4]).

2.2 The concept of exurban areas

We mentioned the present lack of consensus on the definition of exurbanization. However, there have been efforts to narrow the uncertainty of the concept. The exurban area could be defined more specifically using for example population and distance from the central city as a basis, although the exact limits are still likely to be different from one researcher to another. For example, according to Daniels (1999), exurban area is 10 to 50 miles away from a major urban center of at least 500,000 people (zone B in Fig. 3), or 5 to 30 miles from a city of at least 50,000 people. This is generally within 25 minute commuting distance and the population density is generally less than 500 people per square mile. Nelson (1992) on the other hand argues for a definition of exurban counties being those within 50 miles of the boundary of the central city of a Metropolitan Statistical Area (MSA) with a population of between 500,000 and less than 2 million (zone A in Fig. 3), or within 70 miles of the boundary of the central city of an MSA with a population of more than 2 million (zone C in Fig. 3). However, Nelson (1992) points out that it is unclear to determine exurbanization when rural areas become more similar to exurban areas, or when exurban areas become more similar to suburban or urban areas.

The existing definitions of exurbanization of both Daniels and Nelson are summarized in Fig. 3.

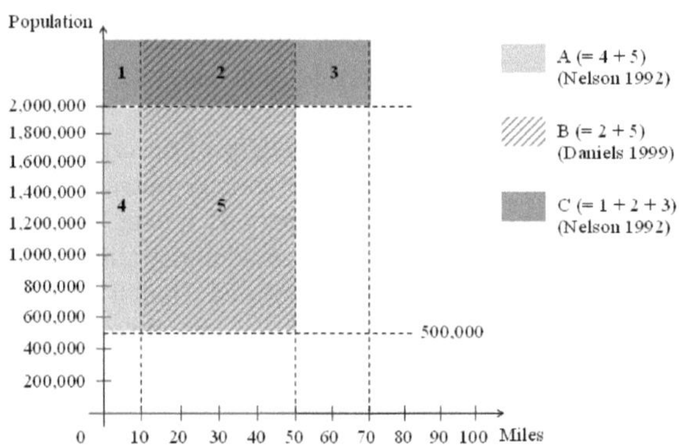

Fig. 3. Differences between existing definitions of exurbanization of Daniels (1999) and Nelson (1992)

In Fig. 3, it is difficult to represent an exact position on the 'Miles' (distance) axis for Nelson's definition since his definition uses the distance from the "boundary of the central city". The extent of central city varies from city to city, and in Fig. 3 we assume the exurban area to be located somewhere between 0 miles and 70 miles but maintain a 50 mile wide band around the city.

These two examples demonstrate that two definitions of exurban areas not only differ in the limits they set for a determining variable such as distance, but also how they use slightly different points of departure for those measurements. It is also apparent that no matter how well defined these urban concepts get in theory, the actual understanding and application is likely to be ambiguous because of their inherently vague character [8]. We therefore propose that these and other terms that will make up future urban ontologies should be defined in a way that explicitly represents their definitional vagueness and makes it possible to evaluate different ontological commitments conceptually as well as spatially. Several approaches have been suggested to represent uncertainty in ontologies. In this work we follow the methodology proposed by Ahlqvist (2005) which is based on the underlying theory of conceptual spaces [10] and that use fuzzy set [31] based formalisms. We formally represent a concept space as a collection, or set, of property definitions. A property definition is represented as a set of values from a certain domain, for example the interval of distance values. The use of fuzzy set based extensions of traditional set theory makes it possible to reconcile the boundaries between different definitions of suburban-exurban and exurban-rural areas acknowledging the graded changes from one zone to another.

3 Exurban areas in Delaware County

Census data is one of the fundamental sources to define the boundary of exurban in terms of calculating total population and population density of unit area. The example dataset consists of block group data of Delaware County, just north of Columbus, Ohio. Population density, the center of Columbus MSA, and urbanized area data all come from U.S. Census2000. Total population of the Columbus MSA was 1,527,948.

The fuzzy membership functions for each definition can be generated by empirical measurement or expert judgment (c.f. [3]). In this case we used the latter approach and elicited membership functions from interpreting exurban areas as a vague concept with the written definitions as a guide to develop membership functions. It is clear that the distance from the center of MSA to each block group increases or decreases linearly. Since the width of transition from one urban zone to another has not been described in the literature yet, we define it arbitrarily for this demonstration. We allow each definition to share fuzzy boundaries with other urban zones. Then we can logically combine multiple memberships, from the distance and population density dimensions, taking the average of the two membership values at any one location and visualize the result in a map.

3.1 Definition of exurban areas according to Daniels (1999)

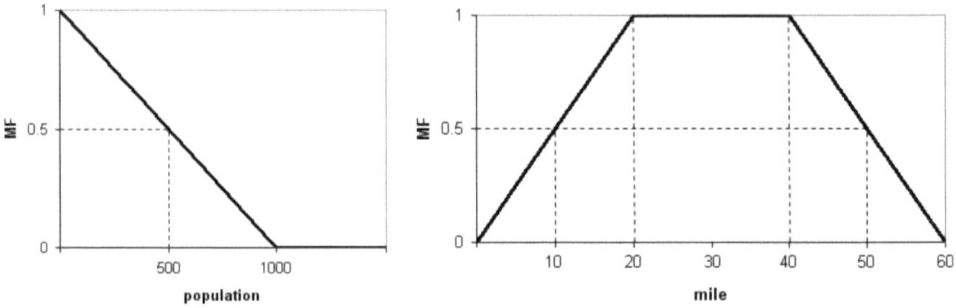

Fig. 4. Membership function of population *(left)* and distance *(right)* in Delaware County based on Daniels's (1999) definition

In Fig. 4 on the left, a simple linear function is used to define the MF value of population. In Fig. 4 on the right, a combination of simple linear functions is assigned to define the MF value of distance. The final membership value for the Daniels definition is determined by calculating the average of the two membership values $MF_{(D)}$ and $MF_{(P)}$.

3.2 Definition of exurban areas according to Nelson (1992)

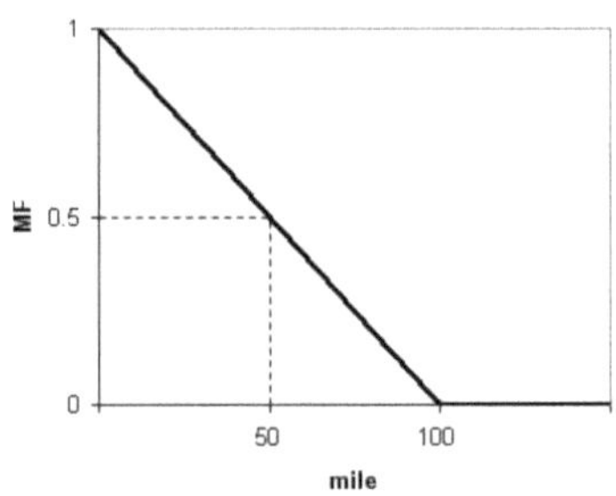

Fig. 5. Membership function of distance in Delaware County based on Nelson's (1992) definition

In Fig. 5, a simple linear function is assigned to define the MF value. In the study area, the whole Delaware County is contained within 50 miles of the boundary of the central city, the City of Columbus. Therefore, the range of distance on the 'Miles' axis for Nelson's definition is assumed to be from 0 to 50 miles in calculating its membership function values in Fig. 5.

4 Results

The maps in Fig. 6 show the difference between crisp membership and fuzzy membership representations of Nelson's (1992) definition of exurban. Since only the distance variable is used in Nelson's area, the map in the left with crisp membership represents the entire area in Delaware County as exurban. However, the map in the right with a fuzzy membership representation shows the gradual transition of MF values of being exurban. From the membership function based on Nelson's definition, areas closer to the center of Columbus MSA have higher MF values of being exurban.

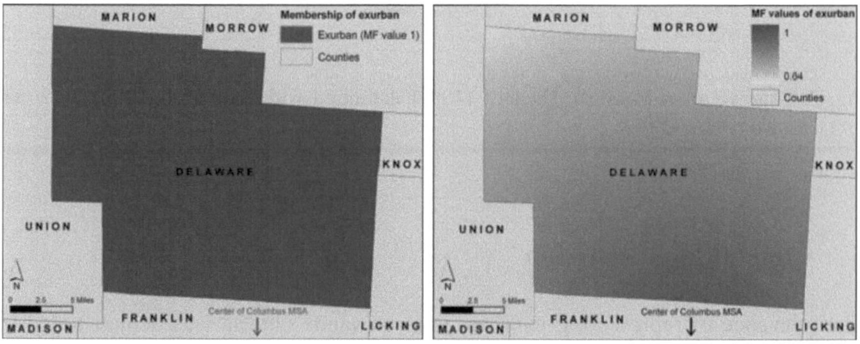

Fig. 6. Exurban areas based on Nelson's (1992) definition with crisp membership *(left)* and fuzzy membership *(right)*

The maps in Fig. 7 show exurban areas with crisp and fuzzy memberships based on Daniels' (1999) definition. In the left map, each block group is assigned to either exurban (MF value 1) or non-exurban (MF value 0). Most of the non-exurban areas are located near central urban areas such as the Columbus MSA in the lower part of the map and near the City of Delaware in the middle-left. In the right map, the block groups have MF values—the degree of being exurban—ranging from 0 to 1 and the MF values are classified into 5 classes for visualization purposes. Since this definition includes distance, the MF values generally increases with distance from the center of MSA. The population component of this definition causes additional variation along the general distance trend. For example, the block groups in the class of values 0.6~0.8 are roughly forming a band around the 10 miles distance from the center of Columbus MSA. Also, some small number of block groups show "leapfrog[1]" pattern in the entire study area. In these two maps, the difference between crisp and fuzzy memberships of the definition is clearly shown. The map with fuzzy membership is able to address a more specific spatial pattern of sprawl in the study area than the map with crisp membership. It shows variations of degree in being exurban among the block groups within the exurban area as well as the non-exurban areas that the map with crisp membership fails to show.

---

[1] If there are discontinuous development pattern s, this is called leapfrog or scattered development (Irwin and Bockstael 2002).

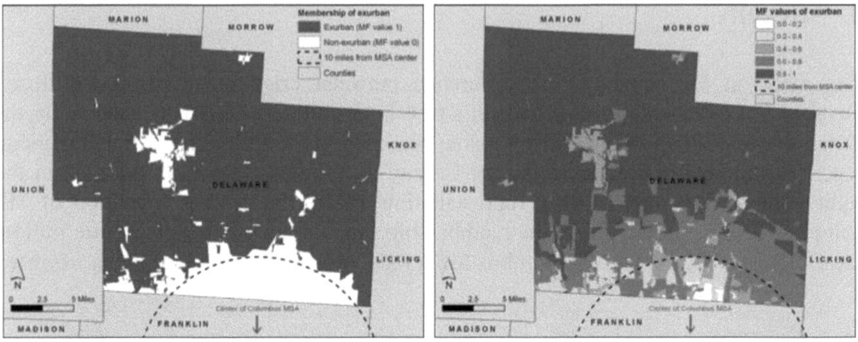

Fig. 7. Exurban areas based on Daniel's (1999) definition with crisp membership *(left)* and fuzzy membership *(right)*

5 Concluding discussion

The relevance of representing exurban areas as vague objects was demonstrated by comparing the traditional crisp representation with a vague, graded representation. The spatial implications of different ontological commitments were compared as they are represented by different definitions of exurban areas. By using the concept of fuzzy ontology in defining exurban boundaries, it is revealed that a clear difference exists between crisp membership and fuzzy membership representations. The crisp classification of exurban area may miss the graded phenomena within such areas.

There are two major differences between our suggested ontology representation and prevailing approaches. First, we propose to generalize the standard first-order logic representation, found in for example the Web Ontology Language (OWL), with a fuzzy-set-based one that can explicitly recognize the vagueness of terms and admit partial belonging to several possible categories. This direction is pointed out by many authors as crucial for developing ontologies to have the expressiveness needed to support practical applications [27];[19]. Although current versions of OWL does not support fuzzy memberships or fuzzy inference per se, we think it is possible to use this or other XML based description languages for fuzzy concept representations, albeit in a not so effective manner. Work on probabilistic and fuzzy extensions [11];[23] of traditional description logics also suggest that it is possible to develop more flexible reasoning capabilities for uncertain ontology semantics. Second, ontology development has this far mostly focused on developing standardized terminologies to support interoperability. This single ontology approach can be contrasted with a hybrid ontology approach [15] in which standardization focuses on the descriptive properties of a term rather than the actual terminology. In our example we can compare different notions of exurban areas by using standard descriptive properties such as population and distance from the city. The combination of these two modifications makes our approach able to compare across heterogeneous terminologies and look for similarities and differences in a flexible manner.

Fisher (1999) argues that ambiguity does come into play in the allocation of social and economic program resources, and it can lead to contention between politicians over the issue of financial support. Urban growth studies of geographic extent, patterns, and classes can be a new resource for local land use decision makers as they plan the future of their communities [29]. In this context, the study of exurbanization showing the uncertainty of its boundaries would also be useful since it reveals the heterogeneity of exurban areas in a location specific context.

This work can be extended in several ways. First of all, we should seek to incorporate the dynamic character of urbanization processes. It is relatively straightforward to visualize the changing exurban boundaries through time in 3D animation of snapshot maps from different points in time. But, we could also further extend the very simplistic category descriptions exemplified here with time dependent characteristics such as time constraints in the definition. This could help identify for example areas where an exurbanization process is just starting. There has been significant work on spatio-temporal modeling using predominantly first-order logic approaches (c.f. [17]). In terms of incorporating dynamic characteristics into our suggested approach, we do not see any major problems to include for example fuzzy time constraints, but this is still a matter of further research. Secondly, a weighted fuzzy membership function can be used for land-use decision making [32]. The subjective weights can lead to large variations [18].

The difference between definitions can be compared by integrating the distribution of fuzzy set values at each location. We might identify satellite nodes within the exurbanization patterns, whose growth is probably related to the central metropolis area and which themselves become secondary sources of exurbanization. Based on these results, such fuzzy membership functions may hence be used to elaborate and test new definitions for exurban areas, in order to better match with the observed empirical phenomenon. We could also investigate the inherent error in the data sets. For example, the Census2000 dataset has confidence intervals in the population density data. By measuring different boundaries using the confidence interval, we can figure out the effect of the error in the data.

References

1. Ahlqvist, O.: Using Uncertain Conceptual Spaces to Translate Between Land Cover Categories. International Journal of Geographical Information Science. 19:7 (2005) 831-857
2. Bishr, Y.: Overcoming the Semantic and Other Barriers to GIS Interoperability. International Journal of Geographic Information Science. 12:4 (1998) 299-314
3. Burrough, P.A., McDonell, R.A.: Principles of Geographical Information Systems. Oxford University Press, New York (1998). Caruso, G.: Integrating Urban Economics and Cellular Automata to model Periurbanisation - Spatial dynamics of residential choice in the presence of neighbourhood externalities. Université catholique de Louvain (2005)
5. Cavailhès, J., Peeters, D., Sekeris, E., Thisse, J. F., 2004. The periurban city.why to live between the city and the countryside. Regional Science and Urban Economics 34 (6), 681–703.
6. Daniels, T.: When City and Country Collide. Island Press, Washington, DC. (1999)
7. Fisher, P.: Ch.13 Models of Uncertainty in Spatial Data. In: Longley, P., Goodchild, M.F., Maquire, D.J., Rhind, D.W. (eds.): Geographical Information Systems. Vol. 2. 2nd edn. John Wiley, New York (1999)

8. Fisher, P.: Sorites Paradox and Vague Geographies. Fuzzy Sets and Systems. 113 (2000) 7-18
9. Fonseca, F., Egenhofer, M., Davis, C., Borges, K.: Ontologies and Knowledge Sharing in Urban GIS. CEUS - Computer, Environment and Urban Systems. 24:3 (2000) 232-251
10. Gärdenfors, P.: Conceptual Spaces. MIT Press, Cambridge (2000)
11. Heinsohn, J.: Probabilistic description logics. In: de Mantara, R.L., Pool, D. (eds.): Proceedings of the 10th Conference on Uncertainty in Artificial Intelligence. (1994) 311-318
12. Irwin, E.G., Bockstael, N.E.: Land Use Externalities, Growth Management Policies, and Urban Sprawl. Regional Science and Urban Economics. 34:6 (2004) 705-725
13. Kemp, K., Vckovski, A.: Towards an Ontology of Fields. In: Third International Conference on GeoComputation, Bristol (1998)
14. Leyk, S., Boesch, R., Weibel, R.: A Conceptual Framework for Uncertainty Investigation in Map-based Land Cover Change Modelling. Transactions in GIS. 9:3 (2005) 291-322
15. Lutz, M., Klien, E.: Ontology-based retrieval of geographic information. International Journal of Geographical Information Science. 20:3 (2006) 233-260
16. Nelson, A.C.: Characterizing Exurbia. Journal of Planning Literature. 6:4 (1992) 350-368
17. Peuquet, D.: Representations of Space and Time. Guilford Press, New York (2002)
18. Robinson, V.B.: A Perspective on the Fundamentals of Fuzzy Sets and Their Use in Geographic Information Systems. Transactions in GIS. 7:1 (2003) 3-30
19. Sheth, A., Ramakrishnan, C., Thomas, C.: Semantics for the Semantic Web: The Implicit, the Formal and the Powerful. International Journal on Semantic Web & Information Systems. 1:1 (2005) 1-18
20. Smith, B.: On Drawing Lines on a Map. In Frank, A., Kuhn, W., Mark, D.M. (eds.): Spatial Information Theory - COSIT 95. Vienna (1995) 475-484
21. Smith, B.: An Introduction to Ontology. In: Peuquet, D., Smith, B., Brogaard, B. (eds.): The Ontology of Fields. NCGIA, Bar Harbor (1998) 10-14
22. Smith, B., Mark, D.: Ontology and Geographic Kinds. In: International Symposium on Spatial Data Handling. Vancouver (1998) 308-320
23. Straccia, U.: Reasoning within Fuzzy Description Logics. Journal of Artificial Intelligence Research. 14 (2001) 137-166
24. Teller, J., Keita, A.K., Roussey, C., Laurini, R.: Urban Ontologies for an Improved Communication in Urban Civil Engineering Projects. In: Proceedings of the International Conference on Spatial Analysis and GEOmatics. Research & Developments, SAGEO, Avignon (2005)
25. Theobald, D.M.: Landscape Patterns of Exurban Growth in the USA from 1980 to 2020. Ecology and Society. 10:1 (2005) 32
26. U.S. Census Bureau: Census 2000 Summary File 1 Technical Documentation. U.S. Census Bureau, Washington, D.C. (2001)
27. Uschold, M.: Where are the semantics in the semantic web?. AI Magazine. 24 (2003) 25-36
28. Wiederhold, G., Jannink, J.: Composing Diverse Ontologies. In: 8th Working Conference on Database Semantics (DS-8), Rotorua (1999)
29. Wilson, E.H., Hurd, J.D., Civco, D.L., Prisloe, M.P., Arnold, C.: Development of a Geospatial Model to Quantify, Describe and Map Urban Growth. Remote Sensing of Environment. 86 (2003) 275-285
30. Wolman, H., Galster, G., Hanson, R., Ratcliffe, M., Furdell, K., Sarzymski, A.: The Fundamental Challenge in Measuring Sprawl: Which Land Should be Considered?. The Professional Geographers. 57:1 (2005) 94-105
31. Zadeh, L.A.: Fuzzy sets. Information and Control. 8:3 (1965) 338-353
32. Zeng, T.Q., Zhou, Q.: Optimal Spatial Decision Making Using GIS: A Prototype of a Real Estate Geographical Information System (REGIS). International Journal of Geographical Science. 15 (2001) 307-321

Preliminary insights on continuity and evolution of concepts for the development of an urban morphological process ontology

Eduardo Camacho-Hübner, François Golay

Laboratoire de systèmes d'information géographique
Ecole Polytechnique Fédérale de Lausanne,
1015 Lausanne, Switzerland.
{Eduardo.Camacho-Huebner, Francois.Golay}@epfl.ch

Abstract. The depiction of urban morphological processes needs the construction of a specific ontology based on a double temporal approach in which synchronic and diachronic relationships must cohabitate. The aim of this paper is to discuss continuity matters of this kind of ontologies and, especially, the effective capability of concepts to evolve as our perception of the complexity of processes grows.

Keywords: Morphology, Process, Ontology, Continuity

1. Introduction

This preliminary paper aims to explore the main contributions of the ontological approach to the depiction of a semantic model of urban morphological processes. We will show the general background of our approach; the complexity of the models embedded on historical analysis, several matters on continuity of concepts frequently used to describe morphological processes and some conceptual problems inherent to both temporal approaches, diachronic and synchronic, needed to uncover the global complexity of urban processes.

Before entering the main theme of this work, it seems relevant to give some information about the general theory called *urban morphology*. Under the concept of urban form, one can find many different meanings. This polysemy can be found at least in three different approaches leading to the study of urban shape: urban growth and urban morphologies using formal models and techniques such as fractals or cellular automata [1],[2],[3],[4]; space syntax, which derives from a set of analytic measures of configurations based on the observation of human behaviour such as people movement in a given urban environment [5], [6]; and the historic-geographical approach, led primarily by Conzen [7] and Whitehand [8], [9] in Britain and combined with the architectural typomorphology studies led by Muratori and Caniggia in Italy [10], [11], [12], [13], [14], [15], and Panerai and Castex in France

E. Camacho-Hübner and F. Golay: *Preliminary insights on continuity and evolution of concepts for the development of an urban morphological process ontology,* Studies in Computational Intelligence (SCI) **61**, 95–107 (2007)
www.springerlink.com

[16]. The present work is embedded in the last two schools of urban morphology attempting to formalize this multi-scale analysis of urban form.

Urban morphology is an explanatory theory interested in the study of the physical form of the city, the progressive constitution of the urban fabric and the analysis of the reciprocal relationships between the constitutive elements of the urban fabric defining particular combinations of spatial features (squares, public spaces, etc.) [17]. In this theory, cities are seen as a composite of cultural, anthropogenic and geographical objects interacting with each other and being able to be "read" in the depth of history and at a given scale. These objects are called urban forms. Within this theory, our on-going project focuses on the study of the changes of these urban forms. Thus, the evolutionary processes of forms identified in this theory are based on the understanding of the relationships between physical objects of the city and the sum of many different kinds of actions. In urban morphology, these actions are not simply human actions having an influence on the physical shape of objects. They are a way to explain the changes observed in forms in a given context through time. Here, changes are driven by endogen causes or by the interaction of a form with its environment. In a more formal way, these actions can be stated as an abstraction of the intrinsic relationship coming from agents of change [18]. The nature of these agents of change covers a very large range. They can be seen as living, social, cultural or environmental input over the formal reality. Within these schools of urban morphology, very little work has been done using GIS or other computer-based techniques, mainly to reconstruct past cadastre from archival sources [19] or field campaigns by combining GIS and GPS techniques [20].

We aim to introduce into this historical and interpretive approach a systemic methodology by considering the process as a series of change and by formalizing the processes encountered in the morphological literature by the means of a historical database and an ontology. The starting point to the classification of processes in the urban historical ontology will be developed using the ontology of process given by Sowa [21]. The differentiation of urban morphological processes from more general conceptualizations will help defining both the nature of data and the subsumed concepts needed to describe and study our specific process category. Finally, we will give some perspectives of this exploratory tool for the empowerment of urban analysis.

2. Project framework

Ontologies are used in our project to formalize a historical GIS-based tool, which will help to explore and to unveil processes from archival and cartographic historical data instead of producing new data from predictive formulae. As the data used to build the model come from many different sources (archives, historical maps, old city maps, and GIS cadastral databases) it must first be organized in adequate spatiotemporal data sets. The structure of the database must be built in order to match the needs of historical analysis. Hence, our approach consists of a semantic model of these data (emerging processes are seen as the result of robust systematic relationship explorations between data classes and confirmed afterwards by archival data and historical evidence), it is necessary to give a formal definition of what we call a morphological process, namely a series of signifying physical changes. Therefore, this formalization will form the basis of a historical data-mining procedure. In this development, the first step is to produce ontologies of the temporal and historical meanings of the concept of *process*. Accordingly, in this paper we discuss the formal "deconstruction" of the definition of the urban morphological process in order to illustrate the framework of possible outcomes from our project such as:

- Identification of the most relevant temporal concepts for understanding morphological processes
- Analysis of the evolution between concepts and instances in order to describe the plurality of urban realities throughout history

Dealing with a polysemic terminology from synchronic and diachronic points of view
Analysis of an effective conceptual capability evolving inside a complex ontology
Choosing the best strategy to describe the complexity of urban processes.

We have already identified three main series of changes relevant to our purpose and they are divided in two different categories: changes through time (values and concepts) and changes of perception (meaning). They represent the first stage in the construction of our future ontology. Briefly, let us draw out each dimension using some common examples in urban morphology.

- Change of value: when a concept is stationary through time, the change of the values (instances) becomes the empirical evidence of the existence of a morphological process. For example, the concept of plot - a parcel of land representing a land-use unit defined by boundaries on the ground [22] - is a shared concept in every morphological analysis, but the forms taken by this concept (instantiation) refers to different periods or places, and the evolution of this form is analysed as the process of plot pattern metamorphosis.
- Change of concepts: The evolution of a concept is the hardest kind of process to be identified. For example, a "road" is a quasi-permanent concept and we have many sub-concepts to differentiate a *path* from a *highway*, we also have some forms that do not exist anymore, but that we can still name them using old concepts, and, finally, the most radical change begins when a concept disappears or a new concept emerges. For example, the dissolution of the cadastral partition in the former USSR creates a void in our analysis techniques if we consider urban form as the relationship between buildings and plots.
- Change of meaning: in this process, which is mostly related to the change of the context of enunciation, the sign takes another value and becomes the vector of some different information - for example the scale of representation. In urban morphology, it is usual to study changes at the scale of the plot (tabernizzazione, insulazione, burgage cycles, etc.) and, as well, to study plot transformations at the scale of a neighbourhood (plot pattern metamorphosis, urban fabric, etc.). In this case, we can see that the same concept is used as "container" and as "contents" of the relevant information. This is a good illustration of the application of the "theory of logical types" [23],[24]: we have two different meanings for the same concept depending on the scale of pertinence. This kind of changes, which are not temporal

but show the coexistence of these different meanings, illustrates the idea of polysemy.

In figure 1, we can see the main construction of the model. Data is obtained by extracting information directly from archival and cartographic sources. Each period of time (epoch) feeds a data set defined in the database. To be able to understand this data, it is necessary to structure the urban concepts used at that period in an adequate ontology. This phase is essential, because the knowledge of the city we have nowadays is the result of both the interpretational capacities of the "reader" and the information historical context of the archival sources (data production seen as "contemporary" to the perception of the given historical period).

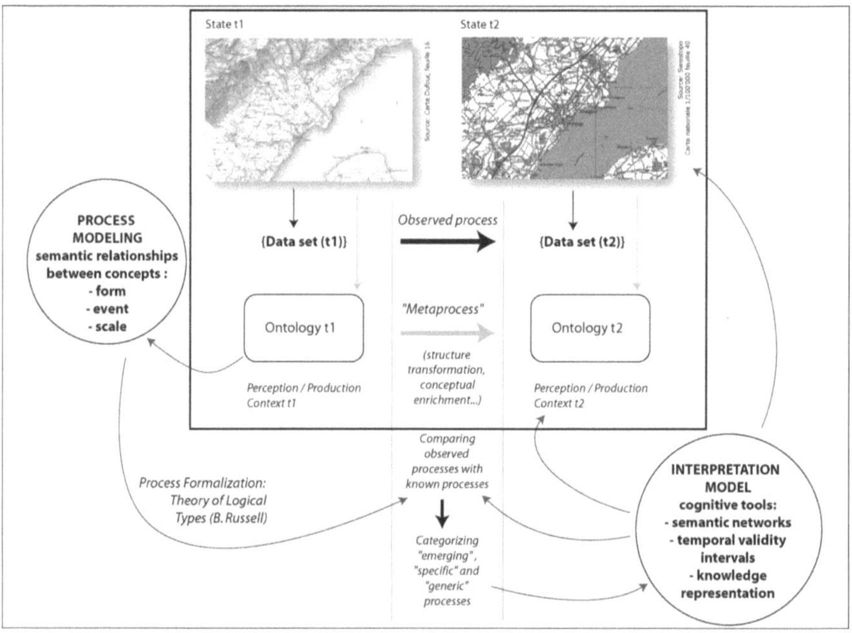

Fig. 1. Interpretation model epoch-based ontologies

This procedure defines the epistemological principle used here: the way reality is perceived in each epoch has an influence on the definition of the whole model. In a first stage, the perception of reality is led by the single-way relationship between entities (reality) and objects (data sets). This relationship is also known as reductionism. In the historical interpretation paradigm that we use, reality is perceived as a whole including *signs* and *meaning of signs through time*. The definition of relationships between entities of the perceived reality and objects in the model needs the modeler to take into account contextual information. This constraint is due to the choice of the interpretation model based on Peirce's *semiosis*, or process of interpretation, in which every phenomenon can be considered as a sign as long as an interpreter refers it to something else [25], i.e., semiosis is a sequence of production of signification by a series of replacements of the sign until its meaning is elucidated inside a given context. The historical context, in which the interpretation is performed, is essential to the comprehension of the relational system between physical features and concepts. Indeed, morphological structures, studied from this historical point of view, are meaningful because of their representation, and because of the "act of utterance" or projection onto a logical space. This logical space is built from a network of morphological concepts. Therefore, one of the ontologies we are trying to build is a model of the historical context enabling the interpretation of the physical data by referring it to morphological and historical evidences. By doing this work, it is possible to study processes as a series of changes observed in temporal data if and only if the definition of these data are in conformity with the conceptual status of these processes at a given period of time. For example, if the concept of cadastral partitions (plots) has the same conceptual value from one period to the next, then this will lead to understanding changes of instances as a process called "plot pattern metamorphosis" [22]. To achieve this objective it is necessary to define simultaneously the data set, directly from the sources, and the concept leading to the description of a given process (ontology).

Finally, in order to give a complete overview of the different implications of this formalization, we would consider the evolution of concepts from one ontology to the next as another kind of process. We call this evolution a "metaprocess" to differentiate it from the data processes described above. This point will be discussed later on.

3. Process-based modeling

In the field of urban morphology as discussed here, a morphological process is understood as a set of changes shaping urban form. These include adaptive, additive, repletive and transformative processes [26]. To adapt this definition to our purpose, it is necessary to show in a few steps the complexity hidden behind its apparent simplicity. Figure 2 illustrates the hierarchy of conceptual complexity starting from a very simple process (plot split) to a complex one (redevelopment cycle [27]).

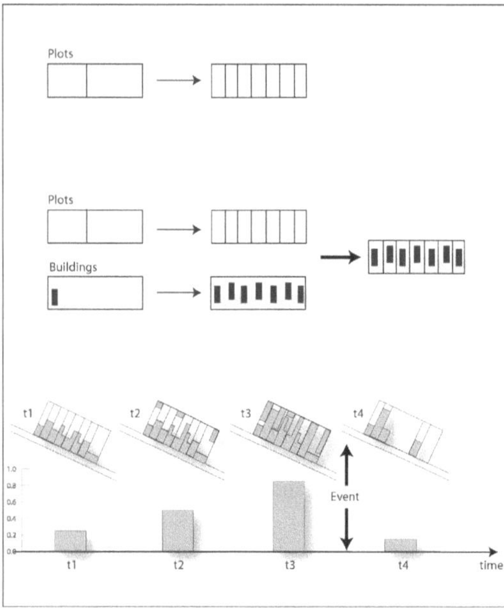

Process name	Complexity Degree	Nature of the Process
Plot Pattern Metamorphosis (Conzen, 1969)	Simple Process	Series of Geometrical Transformations and Topological Relationships. **Monodimensional Semantics**
Building Repletion (Conzen, 1981)	Semi-complex Process	Series of Geometrical Transformations and Topological Relationships. **Multidimensional Semantics**
Redevelopment Cycle (Conzen, 1960)	Complex Process	Series of Geometrical Transformations, Topological Relationships and Event-based Transformation. **Spatial AND Temporal Semantics**

Fig. 2. Example of growing complexity of morphological processes

- The first level of complexity can be defined as a single series of geometrical transformations or topological relationships: division of a plot in several sub-plots, which defines a new *plot pattern*
- The next step combines different simple processes including several series of geometrical transformations, each series concerning a single class of objects for example, *building repletion* (plot split and building appearance). To describe this kind of processes it is necessary to combine several elementary processes. In this case, the semantic relationship is a topological one, building appearance *inside* a new plot pattern
- Finally, the evolution of this *building repletion* through time can be radically modified by an exogenous cause or event. This cause, which is historically known, induces a change in the urban form; in our example, this change is produced by a law adopted for *urban slums clearances*.

In brief, the series of changes observed in the urban shape are all defined by an action, which can be generalized using the concept of *event*. For example, in the geometrical transformation case, the study of economic exchanges helps understanding the evolution of some plot patterns [28]. The importance of the *scale* in this kind of analysis is determinant, because everything could be understood as an event, depending on the *coarse grain* (spatially and/or temporally) of the complexity in which we are interested. Thus, from a *micro-historical* point of view, legacy related to the death of the owner, or trade can be interpreted as an event. In the same way, the *Public Health Acts* that rule all the hygiene problems in British towns since the 19th century can also be read as a series of *macro-historical* events leading to the understanding of the evolution of urban forms. The study of complex processes needs a conceptual construction robust enough to be applicable to many different scales.

In morphological terms, these changes in time and space have been studied in one of its most influential theories [29]. This theory examines transformations due to local adaptations of the way of life i.e., vernacular changes and also much more generalized transformations i.e., codified changes. The vernacular changes occur every time that a building operator finds himself working in continuity with the inherited cultural experiences, also called *spontaneous consciousness,* whereas codified transformations of the built landscape, also called *critical consciousness*, are put in discussion every time systematic experience leads the building operator to choose among different existing alternatives. But even in this second option, there is no certainty on the use of a single "know how" method. One of these concepts is what we introduced above as an *event*. Events are defined by their temporal nature (morphological processes are changes of form over *time* at a given scale) and by the kind of influence they have over forms. The characterization of events is due to the causal nature of such an occurrence in time. On one hand, there are "*morphological events*", which have a role in the changes of form. On the other hand, there are those used to define the context, which means to give clues to explain the occurred event. Considered that the contextual information helps locating historical knowledge, we call the events belonging to this second sub-class "*historical events*". Inside the morphological events sub-class, the temporal scale is needed to define the duration of those events explaining the changes. In the historical events sub-class, the temporal scale is only needed to place historical events in the arrow of time, respecting the temporal semantics of what is called the ordinal sequence of a given process.

4. Groundwork for an Urban Historical Ontology

To summarize our work aiming to define an urban historical ontology needed to explore the semantic network of morphological processes, we can already define three different levels:

- *Historical context ontology*: bottom-up ontology built from partial data sets (Figure 3) in order to interpret historical data.
- *Morphological processes ontology*: based on Sowa's process ontology in order to arrange data for historical data-mining.
- *Time ontology*: using Leibniz and Newton conceptualizations of time. The former to analyze time as the intrinsic property of processes i.e., as a *relationship* between meanings of the same concept from one historical period to the next in order to analyze their evolution and the latter to formalize the notion of *event* mentioned before showing how extrinsic actions helps building a meta-model of ontologies which can be understood as morphological processes.

Due to the fact that historical databases have only partial data sets (Figure 3), temporal ontologies, built from existing and incomplete knowledge of the past, can be identified as a series of *synchronic cuts*.

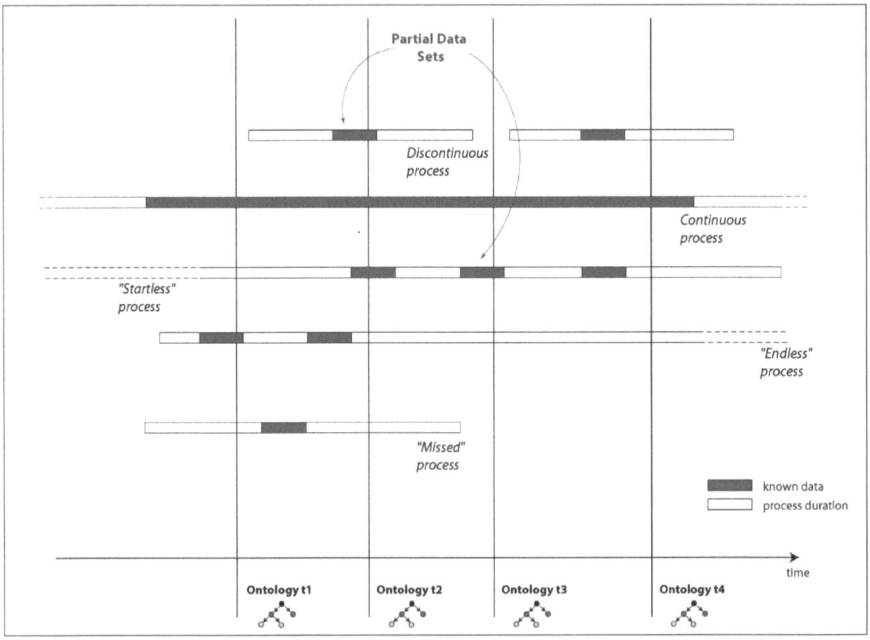

Fig. 3. Synchronic ontologies and diachronic processes

The model of our database does not aim to produce any data but instead tries to understand the processes unveiled by the chunks of coarse information. Therefore, in the case of the historical interpretation, data is only known from the research on archives (primary sources) or by the interpretation of historical works (secondary sources), so they can only represent discrete processes. "Real" processes, which are a combination of economic, social, cultural and environmental causes, are in fact continuous. In consequence, it is not possible to reconstruct complete series of data that are cyclic or stable enough to reproduce results of the same phenomena through reliable intervals of time. This is a fundamental problem of the interpretive approach. One can make as many different epoch ontologies as desired, but it is impossible to synthesize the complexity of continuous process through the definition of ontologies from synchronic cuts. Indeed, as processes are only detectable by the results they produce over the urban form, represented by data and interpreted as morphological evidence, the probability of missing processes is high. We have also the intrinsic difficulty of defining the "starting point" of a given process or the "end" of another one. This differentiation of processes from their continuity temporal attributes is not discussed any further but hints at future developments of this research work.

It is important to point out that in our approach, continuity of concepts needs to be related with sequential time concepts. So it is necessary to use the GIS technology, not as a model from the deterministic point of view, which means that it cannot predict changes of urban form over time, but as a tool for transforming data helping urban morphologists to access to an upper synthesis of data, which can hence unveil new or already known processes when studying a given place.

The interpretation model applied to the historical approach needs then the exploration of time as a *sequence* as well as a *pointer* in history, which means that we need to include both classical temporal concepts [30] in our ontology of urban morphological processes. These concepts can be summarized as follows:

- Newton's concept of time as a pointer in the time line is used for the definition of the chronology of historical events and the production of single epoch ontologies. In a more formal way, temporal structure of data is based on the singular position of each *synchronic cut* on the temporal axis in which we can analyze in depth the relationships between all the objects present at that time. This definition of time can be used as a simple but powerful semantic rule to censor anachronisms.
- However, Newton's concept is not sufficient to give a complete temporal framework to describe the temporality of morphological processes. Adding the Leibniz's idea in our construction, which can be roughly synthesized as a sequential time concept, leads to the series of transformations of the urban fabric. The succession of actions induced by this theory completes the definition of the temporal framework of our model.

By doing this "deconstructive" work, we have tried to formalize the structure of the interpretation process needed to make historical data GIS useful. The contribution of the ontological structure helps to indicate that changes of urban forms are reflected in the evolution of objects (geometrical transformations of plots and buildings, among others), and of concepts (enrichment, generalization, gap bridging) linked to the process itself (Figure 4).

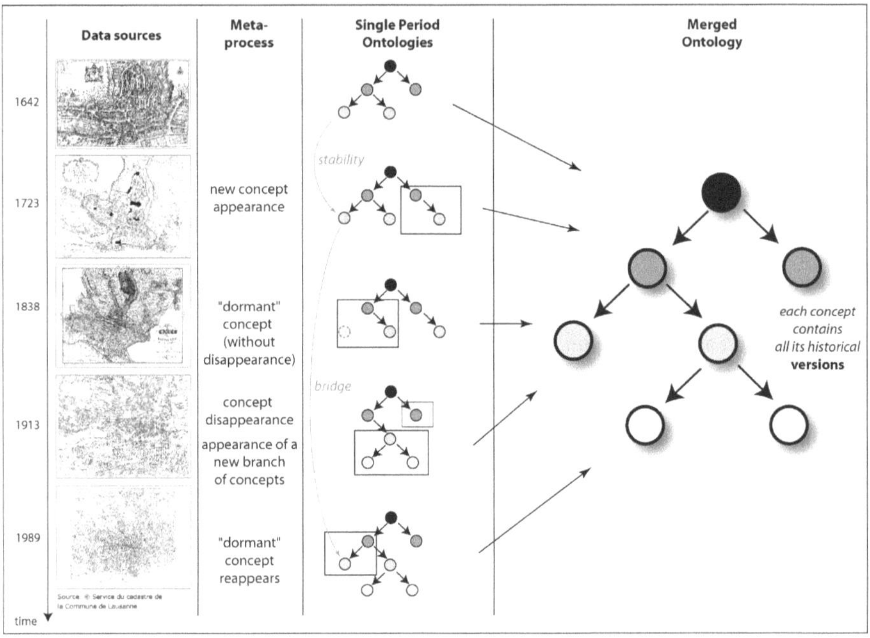

Fig. 4. Concept continuity matters in historical ontologies

From a semantic point of view, the analysis of processes deals with the relationships between concepts in a given epoch, synchronic point of view, and with the evolution of a concept through time, diachronic point of view. In the latter approach the continuity of concepts can be characterized using additional criteria such as: *stability*, *appearance*, *disappearance* and *dormancy* (temporary disappearance of a concept). Finally, the idea of merging all the different epoch ontologies, built from archival and historical cartographic data, provides at the end of this research study a conceptual structure of the city and its evolution through time. To do so, we try to describe not only the form at a given period of time, but the evolution of the conceptualization of the urban configurations through their morphological evidence.

5. Conclusions

This paper discusses an ontology-based theoretical framework for addressing key issues of urban morphological analysis. The main outcomes are summarized as follows:

- The complexity of actual urban processes through time requires a 2-level ontology structure: single period ontologies to address changes of values due to urban change processes, and an overall, merged ontology to address changes of concepts (meta-processes).

- The same urban objects and structures can be considered at different scales, and unveil different processes, changes of meaning.
- Complex urban morphological processes, as Conzen's redevelopment cycles, involve multidimensional spatial and temporal semantics.
- Due to scarce partial data sets usually available on urban history, urban morphologists mainly rely on a series of "synchronic cuts" through urban evolution as a basis for heuristic abduction of relevant diachronic processes.

The proposed framework develops into a genuine, coherent strategy to describe the complexity of urban processes.

Our future research efforts will be directed towards a comprehensive, empirical implementation of the proposed framework within a case study aiming to

- validate and consolidate the proposed theoretical framework and to
- support the inference of relevant urban morphological processes.

However, the implemented prototype should be further developed towards an exploratory platform to support heuristic inductive and abductive reasoning of urban morphologists.

In order to develop the present methodology, it has been assumed that the category of concepts could change over time. This hypothesis is given as an axiom for the improvement of the knowledge about urban processes. The proof of this assumption is not given and it surely will be very difficult to verify its relevance from a formal point of view. Nevertheless, the change of concept as defined in this paper is still necessary to understand the evolution of the perception of the city through history. This fact is of an upper level of abstraction than the idea of *point of view* as currently developed in ontological engineering. From a historical perspective, the comprehension we have nowadays of past processes depends on, at least, three different factors: the conceptual network at the time of the production of the data; the interpretation we give to these data from our contemporary point of view; and, the creation of new categories built today and applied to past configurations in order to seize the complexity of past arrangements.

At this point, it is important not to reduce the complexity of the interpretive process by introducing unquestionable principles, but to give the best description of reality taking into account its multiple temporalities. The discussion about the temporal character of concepts has a very long tradition, since Parmenides and Heraclites to Heidegger. We do not intend to give a final solution to this question, but we agree with the fact that innovation, as well as confusion, comes sometimes with the appearance of the slightest shift inside our shared conceptualizations.

Acknowledgements

This work is supported by research funding from the Secrétariat d'Etat à l'éducation et à la recherche (SER), Département Fédéral de l'Intérieur (Switzerland), Research grant n° C05.0079 "Ontologie des processus morphologiques urbains", and from the Délégué à la Recherche, Vice-présidence aux Affaires Académiques, Ecole

Polytechnique Fédérale de Lausanne (EPFL-VPAA-DAR), Research grant n° 590345 "Conception d'une interface géomatique/morphologie de la ville et du territoire".

We are grateful to the discussions and support from Professor Sylvain Malfroy (*Dictionnaire Historique de la Suisse*), Professor Gilles Falquet, Dr. Claudine Métral and Mathieu Vonlanthen (*University of Geneva*) and to the Direction du patrimoine et des sites (*Geopatrimoine*, Etat de Genève). We would also like to thank both anonymous reviewers who made a number of helpful comments in order to improve this paper. Some of their remarks have already been taken into account, and some others will help us in future development of our project.

References

1. Batty, M. (1991). "Generating urban forms from diffusive growth." Environment and Planning A **23**: 511-544.
2. Batty, M. and P. A. Longley (1986). "The fractal simulation of urban structure." Environment and Planning A **18**: 1143-1179.
3. Batty, M. and P. A. Longley (1987). "Urban shapes as fractals." Area **19**(3): 215-221.
4. Batty, M. and P. A. Longley (1994). Fractal Cities: a geometry of form and function. London, Academic Press.
5. Hillier, B., J. Hanson, et al. (1983). "Space syntax: a different urban perspective." The Architects Journal **178**: 47-67.
6. Penn, A. and A. Turner (2001). Space syntax based agent simulation. 1st International Conference on Pedestrian and Evacuation Dynamics, University of Duisburg, Germany.
7. Conzen, M. P., Ed. (2004). Thinking about urban form: papers on urban morphology by M.R.G. Conzen. Bern, Peter Lang.
8. Whitehand, J. R. W. (1977). "The Basis for an Historic-Geographical Theory of Urban Form." Transactions for the Institute of British Geographers **2**(3): 400-416.
9. Whitehand, J. R. W. (2001). "British urban morphology: the Conzenian tradition." Urban Morphology **5**(2): 103-109.
10. Malfroy, S. (1986). L'approche morphologique de la ville et du territoire. Zürich, ETHZ.
11. Cataldi, G., G. L. Maffei, et al. (1997). "The Italian school of process typology " Urban Morphology **1**(1): 49-50.
12. Cataldi, G., G. L. Maffei, et al. (2002). "Saverio Muratori and the Italian school of planning typology." Urban Morphology **6**(1): 3-14.
13. Marzot, N. (2002). "The study of urban form in Italy." Urban Morphology **6**(2): 59-73.
14. Cataldi, G. (2003). "From Muratori to Caniggia: the origins and development of the Italian school of design typology." Urban Morphology **7**(1): 19-34.
15. Sturani, M. L. (2003). "Urban morphology in the Italian tradition of geographical studies." Urban Morphology **7**(1): 40-42.
16. Darin, M. (1998). "The study of urban form in France." Urban Morphology **2**(2): 63-76.
17. Allain, R. (2004). Morphologie urbaine: géographie, aménagement et architecture de la ville. Paris, Armand Colin.
18. Whitehand, J. R. W. and S. M. Whitehand (1984). "The Physical Fabric of Town Centres: The Agents of Change." Transactions for the Institute of British Geographers **9**(2): 231-247.
19. Koster, E. (1998). "Urban morphology and computers." Urban Morphology **2**(1): 3-7.
20. Lilley, K., C. Lloyd, et al. (2005). "Mapping and analysing medieval built form using GPS and GIS." Urban Morphology **9**(1): 5-15.
21. Sowa, J. F. (2000). Knowledge Representation. Logical, Philosophical, and Computational Foundations, Brooks/Cole.

22.Conzen, M. R. G. (1969). Alnwick, Northumberland: a study in town-plan analysis. London, The Institute of British Geographers.
23.Russell, B. (1908). "Mathematical logic as based on the theory of types." American Journal of Mathematics **30**: 222-262.
24.Kropf, K. S. (2001). "Conceptions of change in the built environment." Urban Morphology **5**(1): 29-42.
25.Everaert-Desmedt, N. (1990). Le processus interprétatif: introduction à la sémiotique de Ch. S. Peirce. Liège, Pierre Mardaga Editeur.
26.Larkham, P. J. and A. N. Jones (1991). A Glossary of Urban Form, Institute of British Geographers.
27.Conzen, M. R. G. (1960). The Plan Analysis of an English City Centre. IGU Symposium in Urban Geography, Lund, C.W.K. Gleerup Publishers.
28.Slater, T. R. (1999). "Geometry and medieval town planning." Urban Morphology **3**(2): 107-111.
29.Caniggia, G. and G. L. Maffei (2001). Architectural composition and building typology: interpreting basic building. Firenze, Alinea Editrice.
30.Thériault, M. and C. Claramunt (1999). "La représentation du temps et des processus dans les SIG: une nécessité pour la recherche interdisciplinaire." Revue internationale de géomatique **9**(1): 67-99.

Theoretical approach to urban ontology: a contribution from urban system analysis

Matteo Caglioni [1], Giovanni A. Rabino[2]

[1] Dipartimento di Ingegneria Civile, Università di Pisa,
via Diotisalvi 2, 56126 Pisa, Italy
matteo.caglioni@lycos.it
[2] DiAP, Dipartimento di Architettura e Pianificazione, Politecnico di Milano,
piazza Leonardo da Vinci 32, 20132 Milano, Italy
giovanni.rabino@polimi.it

Abstract. Building shared and reusable ontologies, both in an operational and more conceptual sense, needs precise definition of system of interest, classification of its relations by means of topological analysis, and explanation of the concepts through mereological tools. The paper presents an attempt to apply these procedures to urban systems, beginning from the corpus of theories developed in urban system analysis to achieve an ontology with the already mentioned suitable features.

Keywords: Building Ontology, Urban System, Semantic Relationships, Topology

1 Introduction

Ontology concept and its use in human sciences are relatively recent, even though this term has been create referring to Aristotle's theories, and in particular it is that philosophy field which is interested in nature and characterization of all what exists [16][21]. This concept has been borrowed by Artificial Intelligence, a field of computer science, in the theory of knowledge, in order to overcome the problem of semantic diversity of information coming from different sources [3][17][20].

However we want to define an ontology, it refers to the concepts of a particular domain of interest, the relationships which connect concepts among themselves and build a structure, and concept definitions which must be done starting from proprieties of (physical, ideal, and social) objects that concepts are referred to, using not a natural language, but a formal one which can be understood by computer [17]. In this paper we refer to guidelines to built ontology provided by Catherine Roussey, in order to define and classify ontology types [16], and we start from previous STSM scientific report of prof. Anssi Joutsiniemi, of Tampere University of Technology, who made his Short Term Scientific Mission at our Department of Architecture and Urban Planning in Milan, where we have discussed about practical and theoretical

M. Caglioni and G.A. Rabino: *Theoretical approach to urban ontology: a contribution from urban system analysis,* Studies in Computational Intelligence (SCI) **61**, 109–119 (2007)
www.springerlink.com

preliminaries of spatially motivated concepts of urban form, in order to open up the conversation and find out common denominators for these issues [11].

Building shared and reusable ontologies, both in an operational and more conceptual sense, needs precise definition of system of interest, classification of its relations by means of topological analysis, and explanation of the concepts through mereological tools (for example decomposition of an object in its parts, or a class in its subclasses). This paper presents an attempt to apply these procedures to urban systems, beginning from the corpus of theories developed in urban system analysis to achieve an ontology of the city with the already mentioned suitable features, underlining in particular three levels (physical, socio-economical, and mental level) through which it's possible to observe the city.

2 Guidelines for urban ontology

We will explain here what we consider as ontologies, how they can be specified, what their significant features are, and how they can be used. We will see that different kinds of knowledge can be distinguished and that knowledge can be modularized in small, manageable pieces. This makes it possible to construct large and complex ontologies out of smaller and more reusable ones.

Our work is based on a bibliographic study of Catherine Roussey, made for COST meeting, in order to provide guidelines for ontology building [16].

2.1 Ontology definition

Artificial Intelligence (AI) has borrowed the word from philosophy and has given its meaning a change. For AI the main question is not what the nature of being is, but what an AI system has to reason about to be able to perform a useful task. Often used and paraphrased definitions of ontology are Gruber's [9] and Studer's ones [18]:

'An ontology is a formal and explicit specification of a shared conceptualization'

A conceptualization is a structured interpretation of a part of the world that people use to think and communicate about the world. In other words conceptualization contains objects, concepts, all other entities that are assumed to exist in a particular area of interest, and all the relationships among them.

In these ontology definition we encounter other really important terms like *explicit* (concept type and their usage constrains are explicitly defined), *formal* (machine understandable), and *shared* (consensual knowledge accepted by group).

2.2 Formal language

In according with Studer's definition [18], we propose to build a formal ontology, using an artificial formally defined language, in order to get as much expressiveness as possible of a natural language, and have the possibility to perform a reasoner on information related with the system to obtain new knowledge.

Objects in ontology have to be defined using their proprieties, which are other concepts linked themselves with other concepts through relationships, in order to built the ontology structure. Natural languages aren't able to describe in a powerful way concept definitions and relationships, which should be represented with another kind of language, more formal.

Currently most of the information is written using syntactical machine readable languages such as HTML. These languages are limited in that they are only intended for human consumption. To fully unlock the potential of such a vast resource of information, we need to make the information not only machine readable but machine-understandable. In order to gain machine understanding we need semantic languages which are able to define meaning to the information being stored. Agents (human or machine) could then use this information in variety of different ways [17].

In order to build an Application Ontology we purpose to adopt Heavyweight characteristics [16], because an ontology with simple taxonomic structure (part-of and kind-of relationships) of concepts, with associated definitions in a natural language, has the same expressiveness of a conceptual map, and it can't be (re)used in other kinds of applications. We have analysed different kinds of formal languages, and we can assert that OWL (Ontology Web Language) seems to be a really suitable language in ontology building.

2.3 Shared and reusable ontology

Ontology building process is characterized by its very high cost and elaborate overlapping activities of development. To build ontology from scratch is too cost-effective. Thus, an approach of ontology construction requires the capture of the key concepts (and their relationships) of a domain. Researchers have proposed many approaches namely bottom-up, top-down, and middle-out. A bottom-up approach for example seems very attractive for many scientific and engineering. The approach focuses on building complex concepts from their primitive (basic) concepts and a list of construction rules.

Research in ontology building from existing ontology sources is motivated by cost and reliability. The recent trend toward ontology library systems to manage, adapt, and control for the purpose of re-use of the great amounts of existing ontologies. Reusability of what exists has proven its success in many areas such as software engineering, medical systems, and environmental information systems. There is, however, limited number of research work in re-use of ontology sources embedded in legacy systems and databases [2].

Starting an urban Geographic Information System (GIS) project presents many challenges. Describing the detail-rich urban environment is one of them. To face this challenge, the use of existing knowledge from previous GIS projects is a necessity. Beyond that, the use of existing data is also desirable. But the lack of formal methods to reuse knowledge and data makes this task really difficult [7].

Reusability can be applied not only to some parts of ontology, or different kinds of data inside that, but we propose to reuse also the reference ontologies, which are used during development time of applications for mutual understanding and explanation between (human or artificial) agents, belonging to different communities in order to establish consensus about concepts [17].

Always in according with Studer's definition [18], an ontology should be shared, and reusability, as we have just seen it, can be an useful tool to obtain this suitable feature. In this work a shared knowledge is a consensual knowledge accepted by a group, and in particular we refer to the corpus of theories developed in urban system analysis, that is at the base of a systemic vision of the city.

2.4 Semantic relationships

The most common way to represent objects in an ontology is through use of semantic relationships between concepts, which give a hierarchical structure to the whole system. The main semantic relationships between concepts are:

- **Taxonomy** (Hiperonomy, Hiponimy): X is a kind of Y (or Y has a kind X)

This relationship is transitive and anti-symmetric, and characterize the relation between classes and sub-classes, where subclasses inherit all proprieties of the their class (i.e. hospital, flat, house are kind of a building).

- **Partonomy** (Meronimy, Olonimy): X is a part of Y (or Y has a part X)

This relationship is transitive and anti-symmetric, and the sum of parts of an object constitute the object itself (i.e. window, door, roof are parts of a house).

Not only semantic relationships are between concepts, but there are also other kind of semantic relationships between verbs:

- **Troponimy** a verb is a troponym of another one, when the first expresses a particular manner of the second (march - walk).
- **Implication** an action implies another one, when the first action can't be performed without to perform also the second (snore - sleep).

Lexical relationships are important relations between concepts that depend by phrases in which they are:

- **Synonymy** two concepts are synonyms, if substituting one concept with the other one inside a phrase, the value of truth of phrase doesn't change.
- **Antinomy** the antonym (or contrary) is a concept having a meaning opposite to that of another concept. A word and its antonym can't be substitute in a phrase, and the negation of antonym preserve the value of truth of the phrase.
- **Polysemy** the polysemous is a concept with more than one meaning.

Semantic relationships are easy to use into an ontology, also because we already know their proprieties and their formal representation, however there are many other kinds of relationships that can be added in an ontology structure, but we need to do an effort to define and characterize them in a formal way.

3 City and urban ontology

Building of urban ontology, proposed in this work, starts form a systemic view of the city: 'city' has seen like a 'machine', a system therefore, modified by man, inside of which he lives, where for living we mean the performance of all those activities characteristic of human being (i.e. eating, sleeping, working, having social relationships, thinking, and having opinions and emotions).

Using this kind of definition, a city can be studied at three observation levels, which represent three different domains, used in urban ontology building: physical level, to which all structures, networks, artefacts on territory belong; socio-economical level, which is related with all activities performed by people into the city and their relationships with other individuals; and mental level pertinent for example with ethics and aesthetics concepts, or with consciousness.

Every one of these levels is important and constitutes three different ways to observe the urban system; in particular the third level is related with the consciousness of the system about itself, referring therefore to scientists' reflections about the city. This consciousness generates mental objects, which have own relationships among them. In detail these objects determine the intentionality of acting, for example architects or town planners design the city in a certain manner, following their idea of beauty, functionality, or optimum.

As previously said, ontology building, in this case the ontology of the city, has to start from a shared knowledge, and for this purpose we have profit of contribute of theory of urban system analysis, and the corpus of territorial methods and models, as informative sources for realization of our urban ontology, moreover we have tried to extract, from these sources, reference ontologies, which were implicit or hidden inside them, and for sure the urban planning scientists have considered them in formulation of their theories.

3.1 Ontology representation

Ontologies could be represented through graphs or diagrams, where objects are punctual elements, and relationships are figured as links or lines which connect different objects.

In according with several philosophers the objects which could be represented through concepts, are three kinds: physical objects that are entities limited in space and in time, social objects that are entities limited only in time (as a contract, or a promise), and ideal objects that are entities not limited in time and space [4][5][6][21].

About relationships, as already said before, the most used ones into ontology representations, are semantic relationships, which give a first hierarchical structure to all considered concepts (they provide a complete and efficacious vision of the used concept catalogue), but we have to say that there are many other types of relationships which can be visualized and which have to be formally defined.

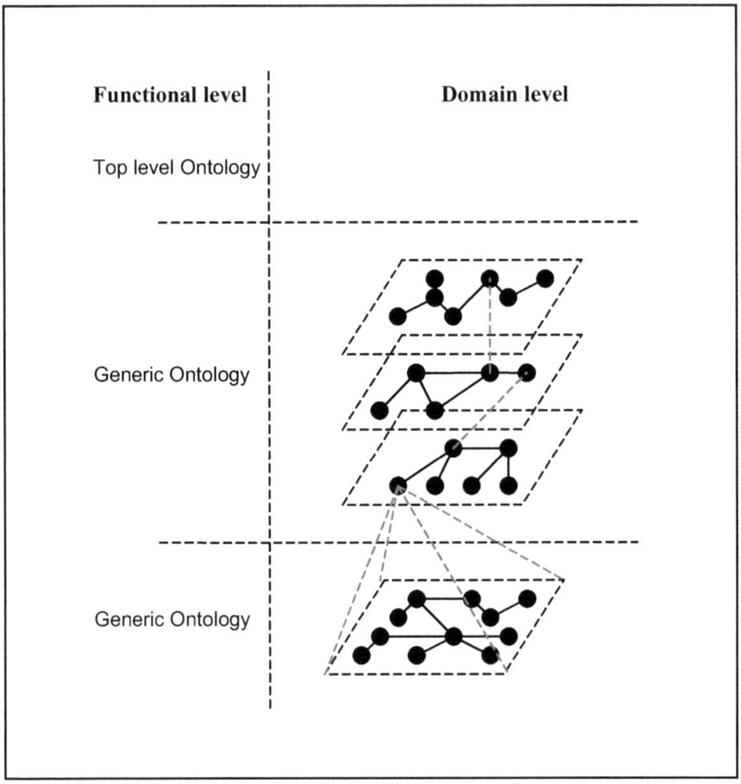

Fig. 1. Concept and relationships representation in an ontology. In this figure is shown also the functional and domain levels, at which we can observe a system

Where Functional level corresponds to a high level abstract view of the operations (functionalities) of the ontology layer. Typically, it is a generic ontology represented by an abstract functional structure consisting of high level ontological concepts and corresponding abstract functional descriptions, which are used to define operations and specify constraints that must be in the domain ontologies. A Domain level consists of one or more domain ontologies that are consistent with the functional level of the ontology layer. Domain ontologies represent the semantics of real world features [2].

Rather than diagrams, ontologies in their entirety should be seen as different layers overlapped, taking into account that an ontology always can be inserted in another one with an higher functional level, or vice versa an ontology can always be decomposed in other smaller ontologies (for mereological difference). Each layer of the same level can be considered like a different visualization of the same concepts: what it changes is the relationships which are represented, and connect in a different way the same nodes (or concepts). Moreover passing from a functional level to another one, it's possible that a concept considered in the higher level, instead, constitutes a set of other concepts and relationships in a lower level.

Regarding visual representation of these ontologies, we need a computerized tool which allows us to choose a kind of relationship rather than another one, in order to analyze how concepts are connected between them, or allows us to see which relationships are available for a particular selected concept. In this direction we have the feeling of a lack of suitable tools.

3.2 A urban system and its ontological representation

The whole set of concepts in our urban ontology can be organized with a hierarchical structure, through the semantic relationships of taxonomy and partonomy, as already explained before, but this is not the only way to arrange concepts. In according with our systemic view of the city, we propose to build an ontology using a classical input-output structure of the urban system. This kind of representation is much more suitable in order to show all mechanisms which act inside the city, and are responsible of its growth: in particular we want to observe and classify all concepts and relationships which determine the so called urban sprawl[1].

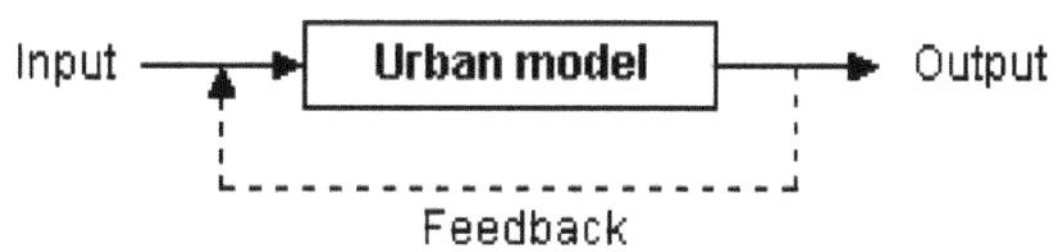

Fig. 2. Input/Output representation of the urban system

The input we can consider in our urban system are for example incoming fluxes of people and vehicles, goods, water, electricity, gas, money, which enter inside a particular model of the city and are transformed into output, like waste, outgoing fluxes of people and vehicles, products, and more important for us, the changes in land use and location of new urbanized areas. Usually this kind of systems isn't linear, but complex, and they have many feedbacks which modify the nature of input.

In this paper, for an easy understanding, we take into account a simplified model of the city: in particular we refer to the Lowry Model [14]. Even if its formulation is rather simple, it provides the relationships between transportation and land use. The core assumption of the Lowry model is that regional and urban growth (or decline) is a function of the expansion (or contraction) of the basic sector. This employment is in turn having impacts on the employment of two other sectors, retail and residential.

---

[1] This issue is based on work made for the Short Term Scientific Mission (STSM) in Finland by Matteo Caglioni and Anssi Joutsiniemi (STSM Host). The results of the mission are discussed in *Caglioni M.: STSM Scientific Report. Unpublished report for Short Term Scientific Mission (Exchange Visits) in the COST C21 Framework (2006).*

- Basic sector. Employment that meets non-local demand. It produces good and services, which are exported outside the urban area. It generates a centripetal flow of capital into the city generating growth and surpluses. Most industrial sector employment is within this category. It is generally assumed that this sector is less constrained by urban location problems since the local market is not the main concern. This consideration is an exogenous element of the Lowry model and must be given.
- Retail sector (non-basic sector). This employment meets the local demand. It does not export any finished goods and services and use the region as its main market area. It accounts mostly for services such as retailing, food and construction. Since this sector strictly serves the local/regional demand, location is an important concern. Employment levels are also assumed to be linked with the local population. This consideration is an endogenous element of the Lowry model.
- Residential sector. The number of residents is related to the number of basic and retail jobs available. The choice of a residential area is also closely linked to the place of work. This consideration is an endogenous element of the Lowry model.

Employment in the basic sector influences the spatial distribution of the population and of service employment. This level of influence is related to transport costs, or the friction of distance. The higher the friction of distance, the closer places of employment (basic and non-basic) and residential areas are.

This model can be described in its mathematical form by the following equations, which represent the core of the whole system:

$$L_i^{tot} = L_i^b + L_i^r$$

L_i^{tot} : total employment for i zone (basic + retail)

$$F_{ij} = \frac{L_i^b \cdot W_i \cdot e^{-\beta d_{ij}}}{\sum_j W_j \cdot e^{-\beta d_{ij}}}$$

F_{ij} : commuting rate from i zone to j zone

$$R_j = \frac{\sum_i L_i^b \cdot W_j \cdot e^{-\beta d_{ij}}}{\sum_j W_j \cdot e^{-\beta d_{ij}}}$$

R_j : employees for each zone

$$P_i = \frac{\sum_j P_j \cdot W_j \cdot e^{-\beta d_{ij}}}{\sum_j W_j \cdot e^{-\beta d_{ij}}}$$

P_i : population for each zone

where W_j is the attraction rate for each zone, and d_{ij} the travel costs.

The Lowry model has obviously several limitations. It is notably a static model, which does not tell anything about the evolution of the transportation/land use system. Furthermore, current economic changes are in the service (non-basic) sectors, forming the foundation of urban productivity and dynamics in many metropolitan areas. A way to overcome this issue is, for example, to consider some non-basic service employment as basic.

Beside the mathematical formalization of the Lowry model, we can describe it through the whole set of concepts and relationships which constitute the urban system. Not only we have a collection of concepts and relations (lightweight ontology), but also we have the possibility to formalize definitions and relationships to build a heavyweight ontology.

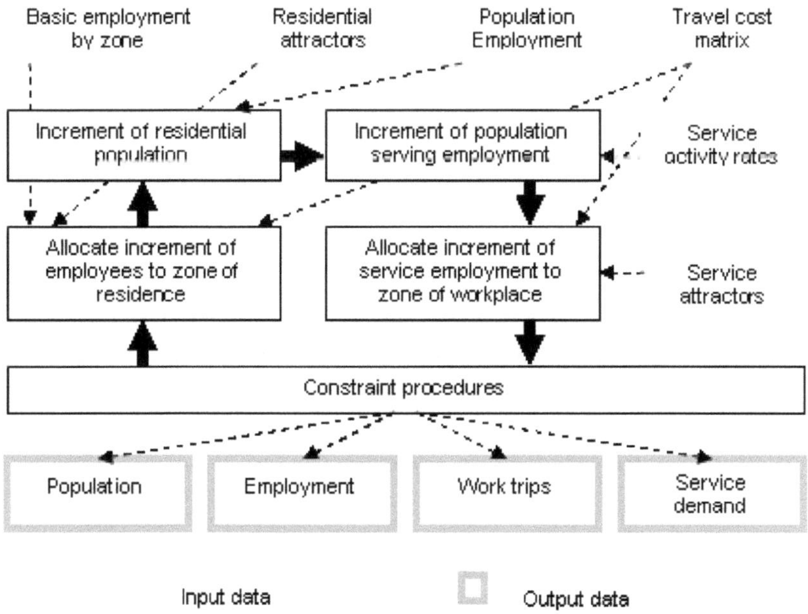

Fig. 3. Ontological representation of the Lowry model

4 Conclusions

In ontology building, it's necessary to begin form a coded set of laws and theories which explicitly or implicitly subtend other reference ontologies. It's quite significant to obtain all the information related with own specific domain, using a kind of shared knowledge. This information provides both concepts and relationships, which should be organized in order to build the ontology structure of the system, and an adequate formal language has to be used. In our case, corpus of urban systems theories, coded and shared, provides us the structure of our ontology of the city.

Once built an ontology of a particular domain, we have the necessity to obtain visual tools which permit us a grater comprehension of the structure of ontology itself.

In particular the representation of concepts and their relationships become really important to understand better the nature of the system itself.

Urban sprawl can be easily classified in an ontology through a taxonomy, but we think that a systemic representation of the concepts and relationships is more suitable for this kind of process. In fact urban sprawl isn't a static concept, but the result of a complex dynamic process acting inside the city. Therefore, to better understand urban sprawl not only we have to study the classes related with this concept, but also we have to formalise all relationships connecting this concept with the other ones inside the urban system domain.

Like we have already said previously, ontology building process is a high cost procedure, above all in term of spent time, so we wish a realization of ontologies which can be reusable, as we already defined it previously, in all those fields that aren't the starting system, which the ontology has been developed for.

References

1. Beck H., Pinto H. S.: Overview of Approach, Methodologies, Standards, and Tools for Ontologies. Third Agricultural Ontology Service (AOS) Workshop, University of Florida, Gainesville, Florida, USA, (2002) p 58.
2. Benslimane D., Arara A., Yetongnon K., Gargouri F., Abdallah H. B.: Two approaches for ontologies building: From-scratch and From existing data sources. In Proceedings of the 2003 International Conference on Information Systems and Engineering (2003).
3. Borst W. N.: Construction of Engineering Ontologies for Knowledge Sharing and Reuse. ISSN: 1381-3617 (CTIT Ph. D-series No. 97-14), Enschede, The Netherlands (1997).
4. Casati R., Smith B., Varzi A. C.: Ontological Tools for Geographic Representation. Published in N. Guarino (ed.), Formal Ontology in Information Systems, Amsterdam: IOS Press, (1998) pp 77–85.
5. Ferraris M.: Lineamenti di una teoria degli oggetti sociali. In A. Bottani, R. Davies (a cura di) L'ontologia della proprietà intellettuale. Aspetti e problemi. Collana Epistemologia, edizioni FrancoAngeli, Milano (2005).
6. Ferraris M.: Ontologia e oggetti sociali. In L. Floridi (a cura di) Linee di Ricerca, Bibliotec@SWIF, Rivista Elettronica di Filosofia, ISSN: 1126-4780, (2003) pp 269-309.
7. Fonseca F., Egenhofer M., Davis C., and Borges K.: Ontologies and Knowledge Sharing in Urban GIS. CEUS - Computer, Environment and Urban Systems 24 (3) (2000) pp 232-251.
8. Fonseca F.: Users, Ontologies and Information Sharing in Urban GIS. In ASPRS Annual Conference, Washington, D.C. (2000).
9. Gruber T. R.: A translation approach to portable ontologies. In Knowledge Acquisition, Volume 5 (2), ISSN:1042-8143, (1993), pp 199-220.
10. Jones C. B., Abdelmoty A. I., Finch D., Fu G., Vaid S.: The SPIRIT Spatial Search Engine: Architecture, Ontologies and Spatial Indexing. In Proceedings of Geographic Information Science: Third International Conference, Adelphi, MD, USA, (2004) pp 125-139.
11. Joutsiniemi A.: STSM Scientific Report. Unpublished report for Short Term Scientific Mission (Exchange Visits) in the COST C21 Framework (2006).
12. Klien E., Probst F.: Requirements for Geospatial Ontology Engineering. In Proceedings of the 8th AGILE Conference on GIScience, Estoril Congress Center, Estoril, Portugal (2005).

13. Lorenz B., Ohlbach H. J., Yang L.: Ontology of Transportation Networks. REWERSE reasoning on the web, Deliverables, A1-D4, (2005) p 49.
14. Lowry I.: A Model of Metropolis. The Rand Corporation, S. Monica, California, U.S.A. (1964).
15. Quine W. V. O.: Ontological Relativity. Columbia University Press, (1977) p 165.
16. Roussay C.: Guidelines to built ontology: A bibliography study. Unpublished COST C21 memorandum (2005) p 16.
17. Smart P. D., Abdelmoty A. I., Jones C. B.: An Evaluation of Geo-Ontology Representation Languages for Supporting Web Retrieval of Geographic Information. In Proceedings of the GIS Research UK 12th Annual Conference, Norwich, UK, (2004) p 175-178.
18. Studer R., Benjamins V. R., Fensel D.: Knowledge Engineering: Principles and Methods. Data Knowl. Eng. 25(1-2), (1998) pp 161-197.
19. Tomai E., Spanaki M.: From ontology design to ontology implementation: A web tool for building geographic ontologies. In Proceedings of the 8th AGILE Conference on GIScience, Estoril Congress Center, Estoril, Portugal (2005).
20. Uitermark H.: Ontology-based geographic data set integration. ISBN 90-365-1617-X, Deventer, The Netherlands (2001).
21. Varzi A. C.: Ontologia. Edizioni Laterza, Roma-Bari (2005) p 178.

A socio-cultural ontology for urban development

Stefan Trausan-Matu

"Politehnica" University of Bucharest
Computer Science Department
313, Splaiul Independentei
Bucharest, Romania
trausan@cs.pub.ro
Romanian Academy Research Institute for Artificial Intelligence
13, Calea 13 Septembrie
Bucharest, Romania
trausan@racai.ro
http://www.racai.ro/~trausan

Abstract. The paper presents an outline of a methodology for developing a socio-cultural ontology starting from Engeström's Activity Theory and the triadic categorization scheme of C.S. Peirce. It also discusses some general ideas on the usage of ontologies in knowledge-based processing. A skeleton of an ontology containing the basic concepts related to the socio-cultural aspects in urban development is introduced. Implementation alternatives are discussed.

Keywords: Ontology, Activity Theory, Urban Development, Knowledge Acquisition, Knowledge-Based Systems

1 Introduction

The growth of towns, combined with the fast cultural changes due to globalization and the population migrations, emphasize the importance of considering socio-cultural aspects of urban development. One desired goal in the near future is the building of the Knowledge-Based Society, in which the already omnipresent computer programs will rather process knowledge that only information. Knowledge processing supposes frameworks that gather the basic concepts or, using a more technical word, the ontology of the considered domain, in order to provide more personalized, more intelligent services. The same ideas determine the evolution of the Web towards a Semantic Web [2], which extends the facilities for knowledge-based processing, collaboration and information retrieval. Ontologies and knowledge processing are also major ingredients of this new generation of the Web.

A socio-cultural ontology for urban development is an essential component if we want to have flexible, extensible, intelligent, knowledge-based programs that can assist urban development specialists to consider socio-cultural aspects in their projects. For example, such an ontology may be used in the semantic search and

S. Trausan-Matu: *A Socio-cultural ontology for urban development,* Studies in Computational Intelligence (SCI) **61**, 121–130 (2007)
www.springerlink.com

combination of web services in urban related applications. It also may be very helpful for the development of natural language processing programs that provide help, answer questions and give advice about, for example, the issues to be considered for further analysis.

The importance of having good ontologies became clear in the knowledge acquisition activities needed in symbolic artificial intelligence programs. However, their success was probably definitively assured in the actual context of information overload due to the expansion of the Web, and in the route to the Semantic Web.

Building an ontology is not a simple activity. It implies philosophical thinking and it is helped if some theoretical outline is provided for the domain for which they are built. For example, John Sowa proposed a top-level ontology [6] starting from the categories introduced by important philosophers (e.g. Aristotle, Kant, Peirce, and others). WordNet, a very successful ontology for natural language processing applications (see, for example, http://wordnet.princeton.edu), was developed starting from psycholinguistic experiments. In the case of urban development, where huge communities of people share buildings, roads, parks, etc., such a theoretical skeleton may be provided by the Activity Theory of Yrjö Engeström [3].

The paper continues with an introduction in ontologies. The third section, after it introduces the Theory of Activity, discusses which could be the basic components of a socio-cultural ontology and how could new concepts be derived.

2 Ontologies

In recent years, the term “ontology” is widely used in computer knowledge-based systems. Probably the most well known definition is the following: “An ontology is a specification of a conceptualization... That is, an ontology is a description (like a formal specification of a program) of the concepts and relationships that can exist for an agent or a community of agents” [4]. Probably one of the reasons of the success of this definition is the fact that it considers several perspectives, going from the computational view to the social particularities of communities.

Ontologies in computer science are represented in computer readable languages (e.g. OWL, see http://www.w3.org/2004/OWL/) as collections of concepts, relations and restrictions. However, their genealogy may be considered from different perspectives: philosophical, computational, and psychological.

Each philosophic system starts with a theory about reality, a theory about what is considered that exists, a so-called ontology. In the process of building an ontology in a computer application, the designer must identify the fundamental categories, the relations and the differences among them. This is exactly one of the main activities that many philosophers, like Aristotle, Kant, Hegel, Peirce also have done. Therefore, philosophy has an important role in ontology engineering. For example, John Sowa, in a knowledge engineering book [6], wrote an entire chapter about the basic ontology he developed, that integrates ideas from the above famous philosophers and from others like Heraclit, Hegel, Leibniz, Whitehead, Husserl and Heidegger. Table 1 shows the basic categories he identified. There is, in fact, nothing surprising. It is normal that an artificial intelligence program has a model of reality and to take into account the very useful work of philosophers in constructing ontologies about reality.

Table 1. Basic categories in Sowa's ontology [6]

	Physical		Abstract	
	Continuant	**Occurrent**	**Continuant**	**Occurrent**
Independent	Object	Process	Schema	Script
Relative	Juncture	Participation	Description	History
Mediating	Structure	Situation	Reason	Purpose

Artificial intelligence aims at developing artifacts able to display an intelligent behavior, similar to that of a human being. Some well-known examples are anthropomorphic robots, expert systems, human language dialogue programs, and, recently, intelligent agents, programs or robots that search information on Internet or give personalized advice to an user.

In many artificial intelligence approaches, computer programs manipulate symbolic structures that represent knowledge, grouped in a so-called knowledge base. Both humans and programs use these structures as intermediates or substitutes for objects in the world. One of the most difficult problems that appear in this kind of systems is the knowledge acquisition problem, which means the process of "filling" the knowledge base with concepts, relations and restrictions. Knowledge acquisition can be facilitated by the existence of an ontology (similarly to a young researcher that much more easily finds his way if he has a sound foundation of basic concepts).

From another point of view, the idea that an artificial intelligence can be achieved via developing a base that includes knowledge that a human has, is directly related to cognitive psychology, which considers that human memory is organized as a semantic network with concepts as nodes and arcs as relations.

Viewing knowledge bases as ontologies has a series of very important advantages for developing knowledge based systems. First of all, an ontology is developed as a coherent framework for the reality and therefore it facilitates knowledge acquisition and machine learning. New concepts may be easily added in such a framework by finding one or some more general concepts and defining some differences between the new concept and the more general ones.

Another advantage is the possibility of developing generic ontologies, including fundamental concepts and to extend such ontologies for every particular application. Finally, ontologies may also enable a friendlier human-computer interaction by tailoring dialog to objects' features from a given context and to users particularities. In fact the ontology of an intelligent program may include not only knowledge about program's environment but also goals, choices, commitments of the program and of the partner (when a dialog is going on).

3 A socio-cultural ontology for urban development

Socio-cultural aspects are major issues in urban development because, of course, towns are built for communities of people. A problem is that communities and collaborative activities cannot be reduced to the sum of individuals. Therefore, for developing a socio-cultural ontology that takes into account all the needed concepts, a theoretic background that considers communities as a basic concept is extremely important. Such a theoretical outline is, in our opinion, Engeström's Activity Theory.

3.1 The Activity Theory

Yrjö Engeström developed his Activity Theory [3] starting from the ideas of the twentieth century Russian school of psychology. The initiator of this prolific sequence of ideas was Lev Vygotsky, which emphasized the role of tools, of words and, in general, of artifacts, as mediators between subjects (humans) and objects [7]. Mediators may be external (physical) or internal (mental) [1]. Therefore, a mediating triangle like that in figure 1 may be identified [7].

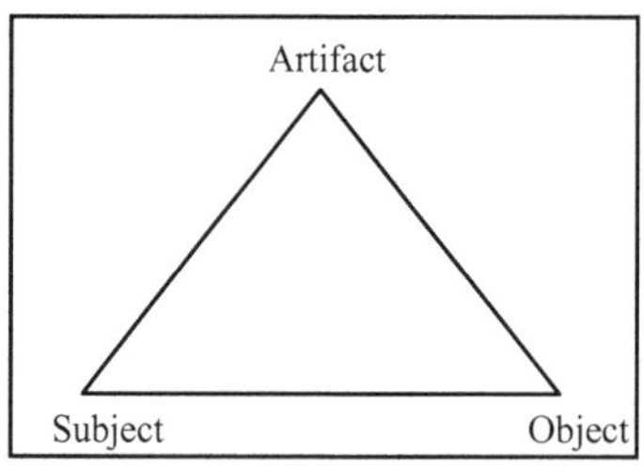

Fig. 1. Vygotsky's mediating triangle

Engeström extended the ideas of Vygotsky by considering the role of communities. Moreover, he identified two new types of mediators: Social rules mediate between subjects and communities, and the division of labor mediates the relation between communities and objects [3]. Vygotsky's triangle is therefore extended with two new triangles, obtaining the diagram from figure 2.

3.2 The basic concepts of the socio-cultural ontology

A problem in ontology development viewed as knowledge acquisition is how to find the methods that identify new concepts and to discriminate among them. Probably the most used method is categorization: identifying the basic categories that encompass a given segment of reality and their particularizations. We will also use this method, the novelty of the paper being its adaptation to the context of the Activity Theory.

John Sowa developed an upper level ontology starting from the basic categories identified by the most famous philosophers, like Aristotle, Kant, Whitehead or Peirce [6]. Similarly, the Activity Theory of Yrjö Engeström provides a theoretical framework that can be used for developing an ontology for urban development that has as basic concepts the two group of entities:

- The three categories: subjects, objects, and communities;
- The three mediators: general artifacts, social rules and division of labor.

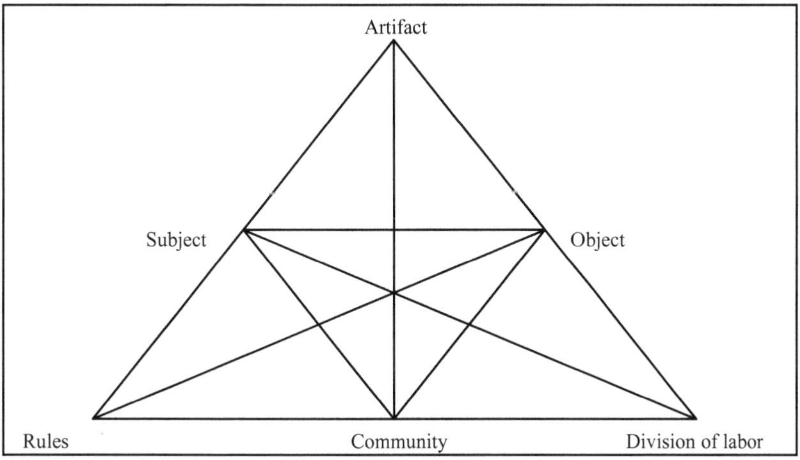

Fig. 2. The activity diagram of Engeström

Each of these six entities will be a basic concept (or "class") in the socio-cultural ontology. These concepts may have attributes, sub-concepts (that may be also sub-concepts of several other concepts, i.e. multiple inheritance of properties is allowed), and relations with other concepts:

- *Subjects* may be classified in several ways, considering different aspects: earnings, social status, ethnicity, age, hobbies, religion, etc. These aspects may be either the basis of a taxonomy of concepts or of attributes. For example, a person that has a habit of walking in a park may either be a new concept, which inherits from the subject concept, or an instance of the subject having "walking in a park" as the habit attribute.
- Different kinds of *objects* may be identified in urban development: buildings, roads, parks, cars, etc. Each of these concepts may be the root of an entire ontology. For example, buildings may be classified in living houses, offices buildings, theaters, cinemas, sport halls, hospitals, factories, shops, etc.
- *Communities* may be classified in the socio-cultural ontology according to several criteria, some of them derived from subjects' attributes like religion or ethnic group.
- *General artifacts* may be physical (tools, objects with a given use, that means that a sub-concept of the object category may be meanwhile a sub-concept of the artifact category), symbolic (texts, prices, taxes) or mental (e.g. imagery, visual patterns, architectural styles).

- *Social rules* may be legislation, traffic rules, unwritten behavior laws or esthetics. Rules may also become artifacts (sub-concepts of the rule category may be also sub-concepts of the artifact category), used by objects in communities.
- *Division of labor* is a basis for the taxonomy of services that assure the functioning and the quality of life of communities (providers of electricity, water and gas, teaching, police, fire department, administration, etc.)

Starting from the three categories “Independent”, “Relative”, and “Mediating” in Table 1, proposed by Sowa starting from Peirce triads [6], we will continue, in the next two sections, the categorization process by considering pairs (relations), and triples (mediators) of concepts.

3.3 Relations

Relations are, together with concepts (classes), the most important ingredients of an ontology. We propose, following the idea of using the Activity Theory diagram, relations to be included in the ontology. For example, below are the most important relations that could be identified in the diagram from figure 2:

subject – object (owned buildings and cars)
subject – rules
subject – community
community – rules
community – object (e.g. buildings, cars, parks)
community – divisions of labor (e.g. roles)
community – artifacts (e.g. beliefs, documents like acts)
object – artifact (property acts, blueprints)
object – subject (owner)
object – rule (of use)

Relations among the vertices of the diagram in figure 2 are extremely important and may be introduced in the socio-cultural ontology in several ways: as attributes, as associations or even as distinct concepts. For example, the relation between a subject and a community may be the “belongs to” attribute of that subject. However, the link of a subject to a rule, for example, may not be a direct link or attribute value. It rather may be derived from other properties of the person (e.g. age may induce that a person has some price reductions). Eventually, the “property” relation joins a subject and an object. Having a distinct concept for this relation allows us to derive a taxonomy of types of property or rent.

3.4 Triples

Triples are not usually explicitly considered in ontologies. However, triples (triads) play an important role in many theories. John Sowa considers them as a fundamental category, following Peirce's ideas [6]. Each of these triples may be seen as articulations, where one concept mediates between other two. As is emphasized in [6], Pierce asserts that triples are the most complex conceptual combinations. Therefore, there is no need to consider quadruples or structures with a higher number of elements.

We will go further with our conceptualization, and, similarly to [5], we will consider different triples that may be identified in figure 2, as suggestions of new concepts for the socio-cultural ontology, that are mediators between other two concepts. Of course, the idea of these mediating concepts might immediately be related with the idea of artifacts.

The implementation of triples in ontologies may be explicit, by including a mediator concept, like in Sowa's ontology [6]. This solution offers the possibility of developing a whole taxonomy of mediators. Another possibility is to have implicit mediators, by including the pairs of mediated concepts in the mediator concept. A third idea is to define a meta-class for mediators.

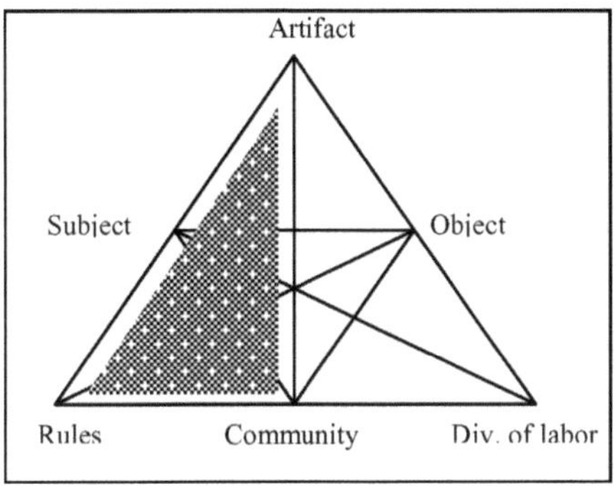

Fig. 3. Image of rules in communities

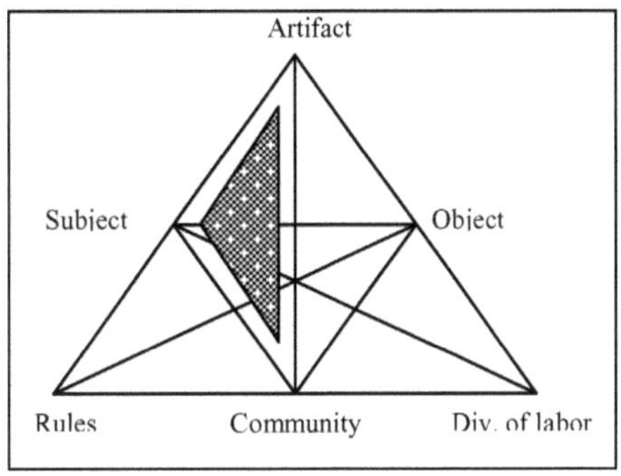

Fig. 4. Artifacts of subjects in communities

The remainder of this section will give several new examples of triples (in addition to those defining the basic mediators: artifact, rules, and division of labor). For example, the triple in figure 3 considers artifacts mediating rules and communities: the mental patterns, symbols or images that some urban or architectural rules establish in communities, like the image of a mountain village induced by the wooden houses. The triple in figure 4 identifies communities' artifacts related to subjects. We can include here history, stories, myths, songs, and collaborative habits.

The triple in figure 5 may be the basic concept for a taxonomy of roles that an individual may play into a community (for example, doctor, professor, priest, etc.). The rules that mediate the access of a community to an object is the concept that is suggested by figure 6.

Figure 7 represents rules (laws) that apply to an individual in relation to an object, for example regarding the property, the rights to modify an object, etc.

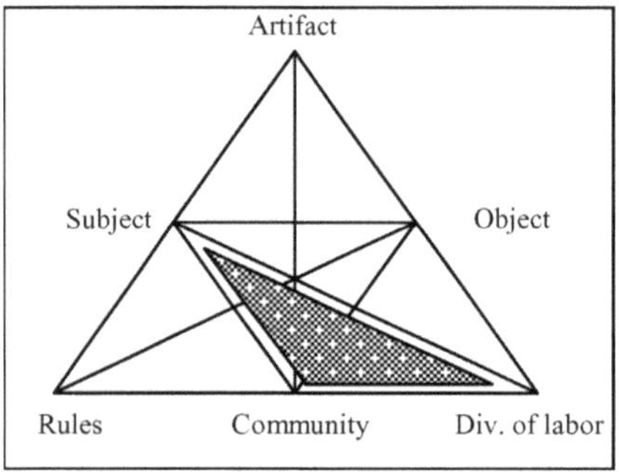

Fig. 5. Roles of individuals in a community

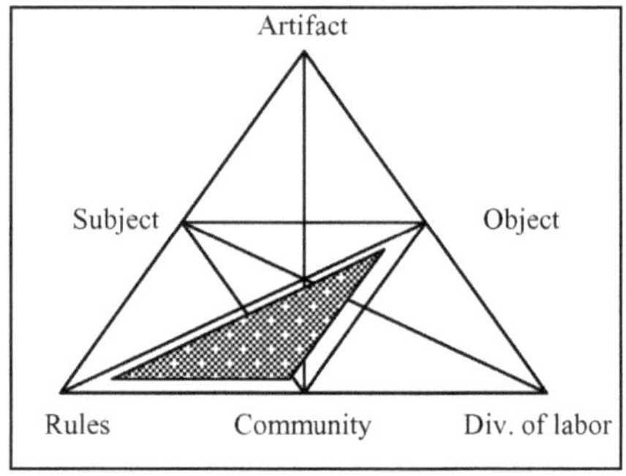

Fig. 6. Rules for objects' use in a community

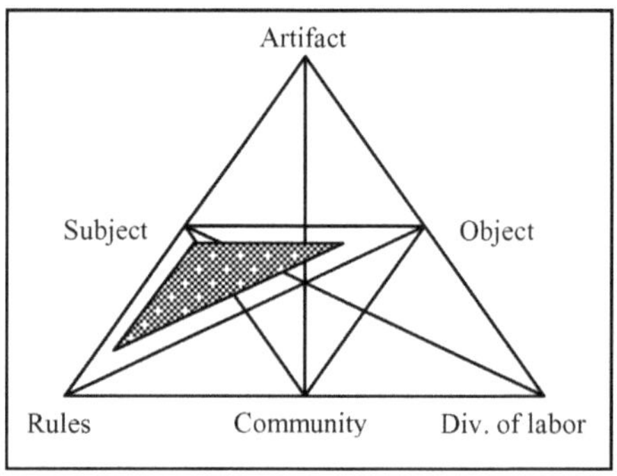

Fig. 7. Rules (laws) that apply to an individual in relation to an object

3.5 An OWL encoding

The basic concepts introduced in the above sections are part of an ontology written in OWL, which was developed and will be further extended in the Towntology COST C21 Action (http://www.towntology.net). For example, a triple of the kind exemplified in fig. 3, and some related concepts are described in fig. 8. The class diagram was generated with the OntoViz tab from Protégé (http://protégé.stanford.edu).

In fig. 9 and fig. 10, the description in OWL of the "mountain_house" class and the "t_community" property are presented (generated also with Protégé). OWL is a standard annotation language based on XML (http://www.w3.org/XML/) and RDF ("Resource Description Framework", http://www.w3.org/RDF/) for describing ontologies that allows definition of classes (concepts), properties and restrictions. The "mountain_house" is a subclass of the artifact_community_rule and has three properties, "t_artifact", "t_rule", and "t_community", the first two having also some attached values.

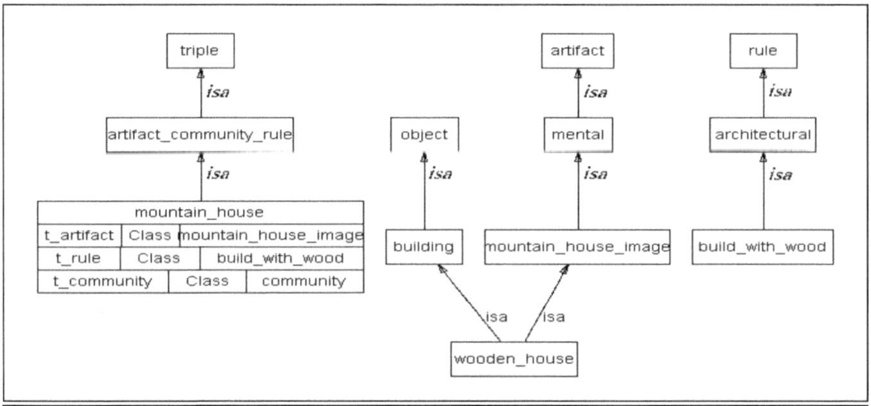

Fig. 8. The "mountain_house" "artifact_community_rule" triple and some related concepts

```
<owl:Class rdf:ID="mountain_house">
  <rdfs:subClassOf rdf:resource="#artifact_community_rule"/>
  <rdfs:subClassOf>
    <owl:Restriction>
      <owl:hasValue rdf:resource="#build_with_wood"/>
      <owl:onProperty>
        <owl:ObjectProperty rdf:ID="t_rule"/>
      </owl:onProperty>
    </owl:Restriction>
  </rdfs:subClassOf>
  <rdfs:subClassOf>
    <owl:Restriction>
      <owl:onProperty>
        <owl:ObjectProperty rdf:ID="t_artifact"/>
      </owl:onProperty>
      <owl:hasValue rdf:resource="#mountain_house_image"/>
    </owl:Restriction>
  </rdfs:subClassOf>
</owl:Class>
```

Fig. 9. The OWL description of the mountain_house concepts

```
<owl:ObjectProperty rdf:ID="t_community">
  <rdfs:domain>
    <owl:Class>
      <owl:unionOf rdf:parseType="Collection">
        <rdf:Description
rdf:about="http://www.w3.org/2002/07/owl#Thing"/>
        <owl:Class rdf:about="#mountain_house"/>
      </owl:unionOf>
    </owl:Class>
  </rdfs:domain>
</owl:ObjectProperty>
```

Fig. 10. The OWL description of the t_community property

The "t_community" property, described in OWL in fig. 10, has as domain (may be used in) the "mountain_house" class, and does not have any attached value.

4 Conclusions

A good ontology in a given domain needs a theoretical framework, that offers some obvious advantages: It facilitates the development of a consistent ontology, it suggests what other concepts should be added, and it prevents the loss of important concepts. We proposed and justified the use of the Activity Theory of Engeström as a framework for developing a socio-cultural ontology for urban development. A first skeleton of the ontology was implemented using Protégé, in the Towntology C21 COST action, and will be extended and experimented in applications.

We have proved in the paper that the usage of John Sowa's methodology of categorization [6], inspired by Peirce triads, and adapted to the Activity Theory [3], generates useful proposals of new concepts to be included in the socio-cultural ontology. However, the resulting relations and triples may be implemented in several ways, the selection of the best variant being not obvious.

References

1. Almeida, A., Roque, L., *Simpler, Better, Faster, Cheaper, Contextual: requirements analysis for a methodological approach to Interaction Systems development,* http://csrc.lse.ac.uk/asp/aspecis/20000003.pdf, retrieved on 23 May 2006.
2. Berners-Lee, T., Hendler, J., and Lassila, O., The Semantic Web, *Scientific American*, May 2001.
3. Engeström, Y. Learning by Expanding: An Activity theoretical approach to developmental research, Orienta-Konsultit Oy, Helsinki, 1987.
4. Gruber, T., What is an Ontology, http://www-ksl.stanford.edu/kst/what-is-an-ontology.html
5. Hewitt, J., An Exploration of Community in a Knowledge Forum Classroom: An Activity System Analysis, in Barab, S., Kling, R., and Gray, J. (Eds.) *Designing for Virtual Communities in the Service of Learning.* Cambridge University Press, 2004.
6. Sowa, J., Knowledge Representation: Logical, Philosophical and Computational Foundations, Brooke Cole Publishing Co., Pacific Grove, CA, 1999.
7. Vygotsky, L.S., 1978, Mind and Society: the Development of Higher Mental Processes, (edited by Cole, M., et. al), Harvard University Press, Cambridge, Massachussets, 1978.

Investigating a Bottom-up Approach for Extracting Ontologies from Urban Databases

Christophe Chaidron [1], Roland Billen [1] and Jacques Teller [2]

[1] Geomatics Unit, University of Liege,
17 Allée du 6-Août, B-4000 Liege, Belgium
{cchaidron,rbillen}@ulg.ac.be
[2] Fonds National de la Recherche Scientifique
LEMA Université de Liège, Lab. of Architectural Methodology
1 Chemin des Chevreuils, B52/3, 4000 Liège, Belgium
jacques.teller@ulg.ac.be

1 Introduction

Ontologies, as "formal and explicit specifications of shared conceptualizations" [1] play a predominant role when developing information systems. This role is increasingly recognised by geo and urban experts when dealing with urban (geo) spatial information systems (GIS, SIS) and spatial databases (SDB). Generally speaking, they provide significant benefits for the design and use of geographic information, such as defining semantics independently of data representation [2]. Urban GIS and SDB are therefore a large source of urban "domain" ontologies [3], like technical networks, urban planning concepts, cadastre structures etc.

It is not yet a common practice to record explicit formalisation of concepts in GIS-SDB documentations. The reason for this is that most GIS-SDB designers have no specific background in ontology design and the role or usefulness of ontologies is still largely underestimated by practitioners. It is hence quite common to have no trace of ontologies in current urban information systems. They are hidden behind documentation, files, database tables or simply part of implicit experts' knowledge. Extracting ontologies from such disparate sources is not a trivial task as it may reveal inconsistencies or gaps in the semantic model underlying these databases.

The aim of this paper is to investigate a *bottom-up approach* for extracting *local ontologies* from urban databases. By local ontologies we mean ontologies related to the databases themselves. Local ontologies of urban SDB contain information about urban phenomena and therefore could be used to (re)construct urban domain ontologies. Different ontology design methods have been presented in the literature, including bottom-up [4] and top-down [5] approaches. A more detailed presentation of such methods can be found in [3]. When dealing with a non-well documented GIS or SDB, this article suggests that starting with defining specifics notions and then extracting more generic concepts by generalisation appears as a pragmatic way to handle ontology generation (extraction).

The paper is organised as follows. First we remind SDB definition and roles of ontology in SDB design. Then we present the empirical bottom-up approach we

C. Chaidron et al.: *Investigating a Bottom-up Approach for Extracting Ontologies from Urban Databases*, Studies in Computational Intelligence (SCI) **61**, 131–141 (2007)
www.springerlink.com

recommended and the next section presents the case study where the approach has been adopted. Finally we draw short conclusions.

2 Spatial databases

There are various sources of information about cities and urban phenomena (plans, maps, registers, etc.). Nowadays, most of the information is stored in numerical format; especially, geographical (spatial) information about urban areas is mostly stored in SDB or GIS. The specificity of these databases is their capacity of storing spatial data, i.e. geographical entities that are described by attributes (standard tuples of a database: alphanumerical data or images, sounds, binary attributes…) associated to some geometric information (position, shape, geometrical and topological relationships, etc…).

Despite an extensive diffusion of such spatial systems and their common use by many citizens, especially through internet (navigation routing, location-based services, visualisation such as "Google Earth", etc.), their conception is not within anybody's reach. System's designers have to follow a formalised methodology, laying stress on the modelling step. More particularly, it requires the creation of specialised documents, according to international standards, like feature catalogue, formalised conceptual data models, and using dedicated tools (Computer-Aided Software Engineering).

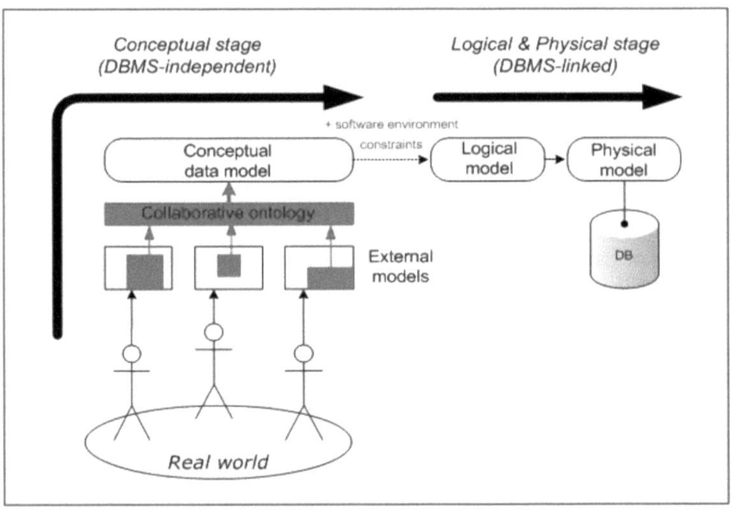

Fig. 1. Classical steps of SDB conception process. (from [6], p. 68, modified)

This stage forces the community of data producers, developers or even future users to (re-)think about the "basic geographic entities of their world" [7], regardless of any database system. It is a natural step in the design process and this usually corresponds to some form conscious or unconscious of domain ontology design (Fig. 1).

Considering ontologies as a necessary step before the creation of tables and relationships is not new [8]. However, formalisation and storage of ontologies is not frequent in SDB and GIS design. It is probably due to underestimation of this highly conceptual stage (lack of knowledge), and the erroneous feeling of "loosing time" when concrete usable results are needed (feature catalogue, CDM, etc.). When urban ontologies have not been explicitly formalised, ontologies can be extracted *a posteriori*, knowing that the database is somehow based on an implicit conceptualisation (inverse order process).

Obviously such an extraction of ontologies from existing databases or GIS is especially relevant in the case of a reengineering of these information systems. As urban information is more and more available in digital format, reengineering is becoming a major concern for most institutions in charge of the maintenance of these data. Data reengineering may indeed be required by the present evolution of techniques (migration from one platform to another one, adoption of open-GIS format), of the requirements (new uses of the databases, increased performance requirements, web access, inter-operability) or the data itself (integration of new information sources, 3D extensions, use of automatic acquisition techniques). In any of these cases, ontology extraction from existing databases and GIS appears as a crucial step before addressing the technical issues of the reengineering process.

3 Spatial objects and relationships

Spatial objects have been formalised for a while (at least in 2D). In SDB and GIS, standardised spatial types are available (such as point, line, polygon, etc.). However, dealing with spatial information is much more than looking to spatial objects; it concerns also spatial relationships existing between them. In this matter, formalisation is far to be finished, even if standards have already been adopted for some type of spatial relationships. In GIScience, one distinguishes between qualitative and quantitative spatial relationships. The former ones do not refer to metrical concepts when the latter ones do. For example, saying that the city of Liège is {*disjoint of, not far from, east of*} the city of Brussels, is a qualitative statement, when saying that the city of Liège is *at 95 km* from the city of Brussels is a quantitative statement. Formalisation of such qualitative concepts is a key research in GIScience. Most of the work in the field has focussed on topological relationships. Such relationships are based on topological geometry and allow distinguishing relations such as "disjoint", "overlap", "included", ... [9] [10]. These are far to be the only qualitative spatial relationships. However, we will restrict our discussion to them in this paper, as they are the only ones to be efficiently managed in SDB and GIS.

Beyond Egenhofer and Clementini operators, there are other ways to express topological relationships. For instance, the formalism CONGOO [6] considers two relations (Superimposition (S), Neighbourhood (N)) with three application levels:

total (t), partial (p), non existent (ne). For example, saying that Liège and Brussels are disjoint could be stated as: Liège S_{ne} N_{ne} Brussels.

This particular way to express topological relationships is equivalent to the ones adopted by the OGC, more information could be found in [6]. We will see that this geo-formalism has been selected for our case study and therefore it is worth mentioning some of its particularity. Beyond the expression of topological relationships, one of the main interests of the CONGOO is to propose the use of *topological matrices*. These matrices contain all the topological relationships that bind the object's sets together. There are two types of topological matrices; the *classical* and the *strong*. The classical matrix contains each topological relationship between every object with all the other objects. The strong matrix contains topological relationships which must exist between a given object and a given number of objects. Figure 2 illustrates the difference between both concepts.

Classical Topological Matrix

	Counties	Municipalities
Counties		-St;-At +Sp,n ; +Ap, n
Municipalities		

As all the counties and all the municipalities are considered, the matrix means that:

- *A county can not be totally superimposed on a municipality*
- *A county can not be totally adjacent to a municipality*
- *A county can be partially superimposed or not superimposed on a municipality*
- *A county ca, be partially adjacent or not adjacent to a municipality*

Strong Topological Matrix

	Counties	Municipalities
Counties		N **+Sp**
Municipalities		

The matrix means that one county must be partially superimposed on N municipalities

Fig. 2. Classical and Strong topological Matrices (from [11])

We do not claim in this paper to summarize topological relationships issues in one section. What is important to note is that such spatial relationships bring crucial information about objects spatial behaviours and consequently about spatial domain ontologies.

4 The bottom-up approach

The proposed bottom-up approach is rather simple and could be theoretically presented as follow (Fig. 3a).

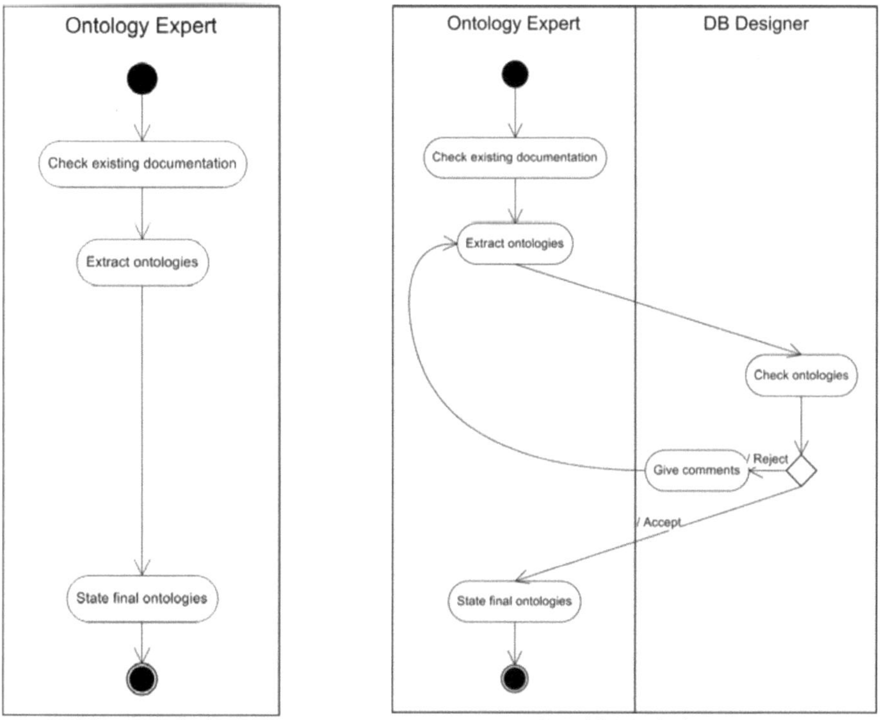

a. Without DB designer b. With DB designer

Fig. 3. Proposed bottom-up approach (UML Activity diagram)

1. The first step is to analyse the existing database documentations and then extract a draft version of the ontologies. Local ontologies can be extracted from data catalogues or data dictionaries and semantic nets can be derived from CDMs (examples of extraction are presented in section 5). The derived ontology should be expressed in an ontology-language like KIF or OWL.
2. At this stage, two options are possible depending on DB designer collaboration.
   a. The relevance of extracted ontologies can be checked by comparing them to the related populated DB. Final ontologies can be then obtained and the extraction process ends.
   b. If it is possible, the next step is to submit the draft ontologies to the DB designer (then the bottom-up approach evolves to figure 3b). An important issue at this stage is to ensure that both "teams" use the same language, the same concepts. A definition is provided for each concept.

This definition includes a textual description as well as a formal expression of its relations with other concepts (IS A, part of and possible topological relations).

3. Remarks formulated by the “DB expert” team must be included in the ontologies extraction process and new ontologies have to be provided until final acceptation.

5 Case study: Brussels UrbIS 2

In Belgium, spatial databases are generally developed by the federal or regional administrations that manage and/or produce inventories of geographic data for the territory they are in charge of. Brussels UrbIS 2 © is the geographic information system of the Regional Government of Brussels.

At the end of the nineties, it became obvious that a complete reengineering of the databases was needed. A collaboration between the *Centre Informatique pour la Région Bruxelloise* (CIRB) and the Geomatics Unit of the University of Liege started in 1998 to provide the necessary support to achieve the reengineering process of part of the SDB (the ADM base containing 33 classes and 830000 instances mostly related to geographical administrative information), i.e. bringing the DB to its second operational version.

The objective of the first conventions was to create *a posteriori* a feature catalogue and conceptual data models. One of the first step was the (re)-definition of local ontologies of the original database [12]. This step has never been formalized for two reasons. Firstly the CIRB team was looking for quick and specific outputs, conceptual stage of the reengineering was not their priority. Secondly ontologies as part of the DB design process were not widely known in the GIS community at that time. Nevertheless, the bottom-up approach we have followed to extract these ontologies can be exposed.

5.1 Application of the bottom-up approach

The practical application of this approach has been rather difficult for several reasons.

First, the existing documentation was incomplete and non standardised. The only documentation available was some relational schemes, a data list (different from a catalogue structure) and data acquisition specifications (for photogrammetric and land surveying acquisitions). The geographical information contained in these schemes was rather poor. Only some hierarchic and thematic links have been deduced from them.

Second, the aim of the work was as we said the creation of DB feature catalogue and CDM, not explicit ontologies. Therefore, the submission process was not based on the validation of ontologies but on validation of these other outputs.

Third, the *database designers* (the CIRB team) failed at the beginning to validate the draft outputs. It was due to a misleading of conceptual perception of the geographical database. Therefore, we had to provide them the necessary tools and methods to formalize their knowledge. It implied to adopt a common language, and more especially a common spatial language. For this purpose we have used first a "natural" language expressed within and Entity/Relationship (E/R) formalism, and later we adopted a more specialized *geo-formalism*. In the nineties, limitations of "traditional" formalisms for handling spatial information were highlighted and consequently several geo-formalisms were proposed: Modul–R [13], GeO - OM [14], MADS [15], CONGOO [6], Geo–UML [16], etc. to name but a few. Overall, these formalisms handle geographic representation of objects as well as spatial relationships. CONGOO has been selected because it was known by the experts in charge of the project.

The practical approach corresponds more to the next diagram (Fig. 4).

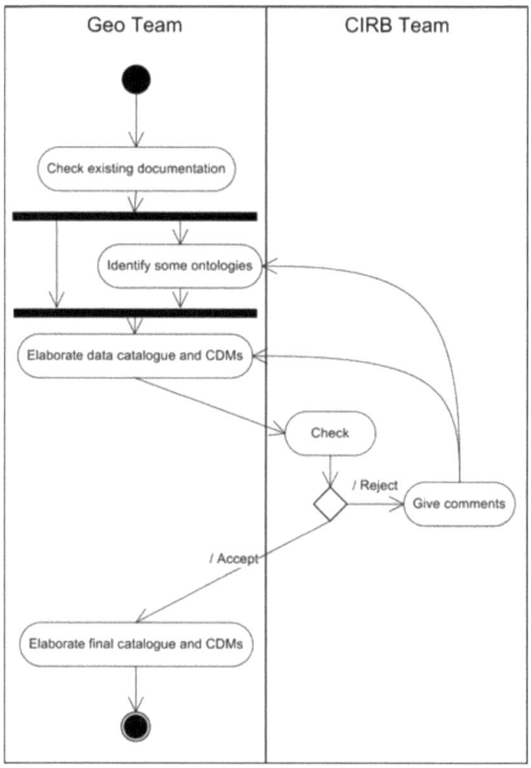

Fig. 4. Practical bottom-up approach (UML Activity diagram)

One of the most important aspects of the submission/acceptation process was the establishment of objects spatial properties: object representation and spatial relationships between objects.

In the reengineering project, topological matrices (*cf.* 3) had to be created with the CIRB team. This is the first step of any CONGOO conceptual model. The elaboration of this matrix is time consuming and a hard step of the process. Quite importantly, by identifying spatial (topological) relationships between objects, this stage revealed object's definition inconsistencies. This point if further developed in the next section (*cf.* 5.2).

5.2 Ontologies extraction and objects definition

The aim of this experience was to define specific notions and then extract more generic concepts by semantic generalisation. The process started with proposing definitions for DB's basic objects, which should be very close to the ontologies that drove the DB's creation. However, one has to keep in mind that it was a reengineering project and therefore we could not ignore the complexity of existing DB's objects. The following example, presenting the evolution of the definition of the "house" object, illustrates the different levels of abstraction we had to consider.

The initial definition was clearly link with object's graphical construction and data sources (in this case the topographical survey).

Definition 1: The « house » is the building extract out of the topographical survey

We did not have the ability to change object's name, however, this "definition" was clearly not satisfactory. Our own understanding of the objects leads us to the following definition (whose validity was checked against other DB's documentation):

Definition 2: The « house » corresponds to footprint of a building (including its annexes)

This definition appeared to correspond to the designer ontologies. However, when considering spatial relationships between objects (from the topological matrix), the definition had to be adapted. The issue was highlighted when considering the topological relationship "superimposed to" (overlap). From objects definitions, it was expected that "house" could not be superimposed to object "street". However, CIRB team indicated that it was indeed possible because of the inclusion of objects such as bus stop, fountain, etc. into the object "house".

Definition 3: The « house » corresponds to building's footprint, including annexes and all other construction such as church, chapel, monument, school, fountain, greenhouse, bus stop, etc.

This definition is quite odd and not satisfactory conceptually. However, it corresponded to the reality of the DB and had been included in the feature catalogue. Of course, one of our DB's reengineering recommendations was to split this object "house" into several more semantically consistent objects.

In this example of the objects definition extraction, we can say that the definition 2 was indeed at a higher level of conceptualization. An ontological dictionary could have been produced at this stage, prior the feature catalogue. The project also raised linguistic issues due to the fact that the ontology had to be developed in both French and Dutch, which are the two official languages in Brussels.

5.3 Conceptual data models and semantic nets

As we have seen above, the extraction of ontologies during the reengineering process was a crucial step in the understanding of objects/concepts. As a logical step in the process, following data cataloguing, conceptual data models were built. One in Entity/Relationship formalism and the other with CONGOO (Fig. 5).

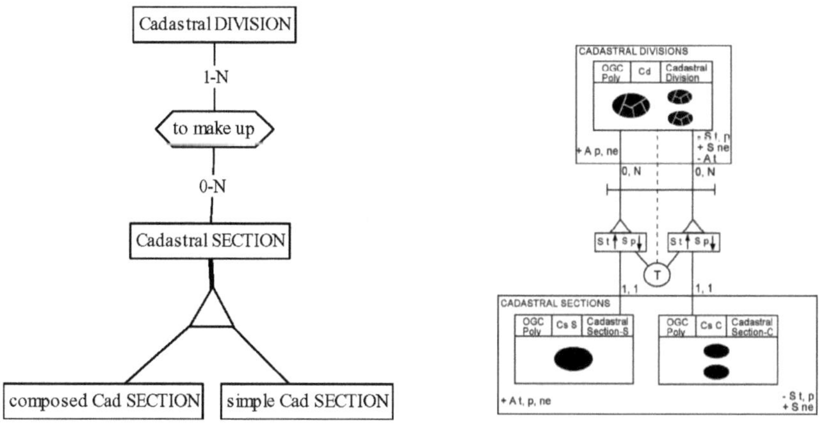

Fig. 5. Extracts of conceptual schemas: E/R and CONGOO

The temptation to (re-)interpret the E/R model as a semantic net is large. If it can become a really interesting and convenient synthesis and communication tool, such a schema is basically designed for a specific information system, describing the contents of a specific database, i.e. the specifications of one possible "world" [17], [12]. That means that we would have to operate an intermediate step to build a kind of semantic net (Fig. 6), based on the generic definitions. By this way, we would obtain a richer model (global-transposable-sharable) than the database conceptual schema, capturing the semantics of information in a formal way, and usable as a possible way for data integration [2]. This extraction process from E/R models can be envisaged (semi)automatically (selection of specific entities, relationships and attributes). It is not the case with CONGOO which is currently only a "graphical" formalism without CASE tools.

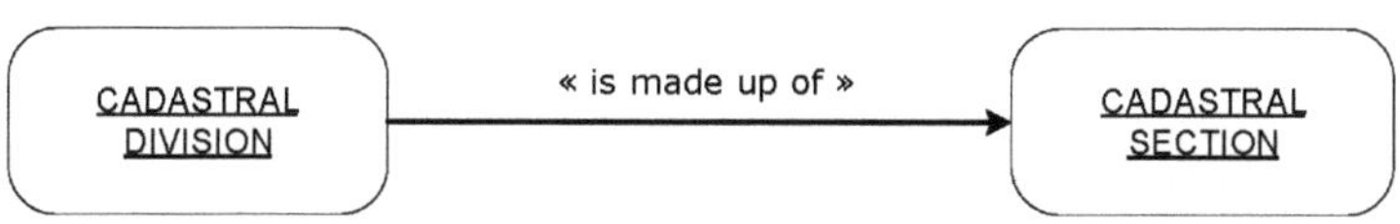

Fig. 6. Semantic net

Two "models" were thus proposed, the E/R model using complex types of relationships and specializing objects based on their geographical representations and the CONGOO model, much richer, including concepts such as classes and layers and representing all topological relationships between objects. It is worth noting that the CIRB team has continued to use and up-date the E/R model (the one closer to the semantic net) and has left behind the CONGOO one as they did not managed to maintain it.

6 Conclusion and future developments

In this article, we have discussed extraction of local SDB ontologies in the context of urban domain ontologies identification. We have tried to clarify the role of ontologies in SDB's design and reengineering. If the ontology level is necessary for DB's design (and interoperability) [18] [19], related ontologies are not always formalized. Therefore, local SDB ontologies are usually hidden in SDBs and associated documentations (feature catalogues and CDMs). In this case, it is possible to extract them from the documentation by applying a bottom-up approach. This process could be improved by a good collaboration with DB's original designer when the DB is poorly documented.

From our experience, extracting local ontologies (and associated objects definitions) implies a very good knowledge of spatial relationships between DB's objects. If extraction processes could be investigated further, it is clear that the major output of this reflective analysis is that local DB ontologies must be recorded during the DB design process. Another issue which should be tackled is the case of non documented DB or more generally non documented spatial numerical information; we believe that a comprehensive analysis of spatial relationships between instances could be the first stage of local ontologies extraction.

References

1. Gruber, T. *A translation approach to portable ontology specifications.* Knowledge Acquisition, 1993**5**(2): pp. 199-220.
2. Morocho, V., L. Pérez-Vidal, and F. Saltor. *Semantic integration on spatial databases.* in *Proceeding of VIII Jornadas de Ingenieria del Software y Bases de Datos.* 2003. Alicante. pp. 603-612.
3. Roussey, C., R. Laurini, C. Beaulieu, Y. Tardy and M. Zimmerman. *Le projet Towntology. Un retour d'expérience pour la construction d'une ontologie urbaine.* Revue Internationale de Géomatique, 2004. 14(2): pp. 217-237.
4. Van der Vet, P.E. *Bottom-up construction of ontologies.* IEEE Transactions on Knowledge and Data Eengineering, 1998. **10**(4): pp. 513-526.
5. Sowa, J. *Top-level ontological categories.* International Journal on Human-Computer Studies, 1995. **43**(5-6): pp. 669-685.
6. Pantazis, D. and J.-P. Donnay. *La conception de SIG. Méthode et formalisme.* 1996. Paris: Hermes, 339 p.

7. Fonseca, F., M.J. Egenhofer, P. Agouris and G. Câmara. *Using ontologies for integrated geographic information systems.* Transactions in GIS, 2002. **6**(3): pp. 231-257.
8. Gruber, T. *The role of common ontology in achieving sharable, reusable knowledge bases.* in *Proceeding of the International Conference on Principles of Knowledge Representation and Reasoning.* 1991. Cambridge. pp. 601-602.
9. Clementini, F., P. Di Felice and P. Van Oosterom, *A small set of formal topological relationships suitable for end-user interaction.* Advances in Spatial Databases LNCS 692, 1993. pp. 277-295.
10. Egenhofer, M. *A model for detailed binary topological relationships.* Geomatica, 1993. **47**(3-4): pp. 261-273.
11. Pasquasy, F., F. Laplanche, Jean-Christophe Sainte and J.-P. Donnay. *MECOSIG adapted to the design of distributed GIS.* in *On the move to Meaningful internet systems 2005*, R. Meersman et al. (Eds): OTM Workshops 2005, LNCS 3762, pp. 1117-1126.
12. Fonseca, F., C. Davis, and G. Câmara. *Bridging ontologies and conceptual schemas in geographic information integration.* GeoInformatica, 2003. **7**(4): pp. 355-378.
13. Caron, C., Y. Bédard and P. Gagnon. *MODUL-R, un formalisme individuel adapté pour les SIRS.*Revue Internationale de Géomatique, 1993. **7**(3), pp. 283-306.
14. Tryfona, N., D. Pfoser and T. Hatzilacaos. *Modelling behavior of geographic objects: an experience with the object modelling technique.* at *CASE*, 1997. Barcelona.
15. Parent, C., S. Spaccapietra, E. Zimányi, E. Donini, C. Plazanet, C. Vangenot, N. Rognon and P.-A. Crausaz. *MADS, modèle conceptuel spatio-temporel.* Revue Internationale de Géomatique, 1997. **7**(3-4): pp. 317-352.
16. Bédard, Y. *Visual modelling of spatial databases: towards spatial PVL and UML.* Geomatica, 1999. **53**(2): pp. 169-186.
17. Bishr, Y.A. and W. Kuhn. *Ontology-based modelling of geospatial information.* presented at *Third AGILE Conference on Geographic Information Science,* 2000. Helsinki.
18. Frank, A. *Spatial Ontology.* in *Spatial and Temporal Reasoning*, O. Stock, Editor. 1997. Dordrecht: Academic Publisher. pp. 135-153.
19. Smith, B. and D. Mark. *Ontology and geographic kinds.* in *Proceedings of the Tenth International Symposium on Spatial Data Handling.*, T Poiker and N. Chrisman, Editors. 1998. Burnaby: Simon Fraser University. pp. 308-320.

Urban Ontologies: the Towntology Prototype towards Case Studies

Chantal Berdier[1], Catherine Roussey[2]

[1]EDU laboratory, INSA of Lyon, bat. Eugene Freyssinet 8 rue des sports 69621 Villeurbanne Cedex France
Chantal.berdier@insa-lyon.fr
[2]Université de Lyon, Lyon, F-69003, France; LIRIS laboratory UMR 5205, Université Lyon 1, batiment Nautibus
43 Boulevard du 11 Novembre 1918, 69622 VILLEURBANNE CEDEX
catherine.roussey@liris.cnrs.fr

Abstract. This article describes our work realised since the beginning of the Towntology project, which aim was the development of urban ontologies. First of all our work deals with the comprehension of what is an ontology and their uses. Thanks to this prior studies, we develop a prototype and 3 urban ontologies: road system, urban mobility, urban renewal. After describing our result, we synthesize the main difficulties encountered during the development of these ontologies.

Keywords: Ontology development, Ontology classification, Towntology prototype, road system ontology, urban mobility ontology, urban renewal ontology

1 Introduction

Since 20 years, we are speaking more and more about ontologies. But what is exactly an ontology? Are ontologies useful for urban planning? What may be the different uses of ontologies in urban planning? This article tries to answer to these questions through a presentation about ontology evolution and case studies.

There are a lot of definitions of ontology notion: philosophical approach, Artificial Intelligence approach, linguistic approach, information retrieval approach, ... In this article, we chose the definition of Artificial Intelligence field proposed by Gruber in 1993: an ontology is "*the specification of conceptualisations, used to help programs and humans share knowledge*" [10].

The *conceptualization* is an abstract, simplified view of the world that has to be representing for some purpose. Thus, the conceptualization result from a modeling choice. The conceptualization determine the universe of discourse that is to say the objects, concepts, and the relationships that hold among them.

The *specification* is the representation of this conceptualization in a concrete form. One step in this specification is the encoding of the conceptualization in a knowledge

C. Berdier and C. Roussey: *Urban Ontologies: the Towntology Prototype towards Case Studies,* Studies in Computational Intelligence (SCI) **61**, 143–155 (2007)
www.springerlink.com

representation language. The goal is to create an agreed-upon vocabulary and semantic structure for exchanging information about that domain.

This article describes our work realised since the beginning of the Towntology project, which aims was the development of urban ontologies. The first section presents a classification of ontology type and their different uses. The second section describes the Towntology prototype towards 3 case studies: road system, urban mobility, urban renewal. Then, we synthesize the main difficulties encountered these cases studies.

2 Ontology Classifications

Researchers's interrogation about expert systems[1] modularity are at the root of ontologies. In fact, these systems were dedicated to the resolution of a particular domain problem without any possibility to evolve to other domain. To resolve this problem of reusability, experts systems has been separated in different units: Problem Solving Method (PSM) unit, domain knowledge unit, matching of domain knowledge on PSM knowledge unit.

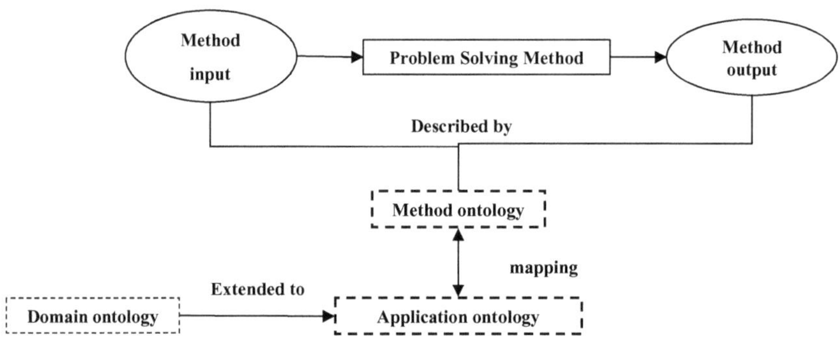

Fig. 1. the architecture of an expert system using ontologies

A Problem Solving Method explains a way to accomplish a generic task. The method ontology specify the input and the output of this method. The domain ontology in the description of the domain knowledge and the application ontology is a limited set of the domain knowledge used for processing the task. Thus, in order to solve a problem for a particular domain, method ontology concepts should be mapped to application ontology concepts [4].

Today ontologies are used in other systems than expert systems. This proliferation have give rise to different categories of ontologies. We distinguish 4 type of ontologies classification[2]. The borderline between each classes is not clearly defined.

---

[1] Expert systems are also called knowledge based systems

[2] these classifications was proposed by [17] and the [11].

1. classification according to formalization,
2. classification according to expressiveness,
3. classification according to purpose,
4. classification according to specificity

2.1 Classification according to Formalization

Four type of ontologies are distinguished, dependent of the language used to represent the ontology.

- highly informal: the ontology is expressed loosely in natural language; The yahoo directory is an example.
- semi-informal: the ontology is expressed in a restricted and structured form of natural language, greatly increasing clarity by reducing ambiguity. A thesaurus like the French architectural thesaurus[3] is a semi informal ontology.
- semi-formal: the ontology is expressed in an artificial formally defined language. The Dublin Core Metadata Initiative [5] which is building a metadata standard in order to describe each resource in RDF format define a semi formal ontology [13].
- rigorously formal: the ontology is meticulously defined with formal semantics, theorems and proofs. An ontology of the standard unit of measurement express in KIF language is a rigorously formal ontology [8].

2.2 Classification according to Expressiveness

This classification is a fusion of preceding classes in two families. The ownership depend if a software can make or not some reasoning process with the ontology.

- Heavyweight Ontology (HO): Heavyweight ontologies are extensively axiomatized and thus represent constraints explicitly. The purpose of the axiomatization is to exclude terminological and conceptual ambiguities, due to unintended interpretations. Every heavyweight ontology can have a lightweight version. Many domain ontologies are heavyweight because they should support heavy reasoning (e.g., for integrating database schemata, or find some solution of a particular problem like medical diagnosis).
- Lightweight Ontology (LO): Lightweight ontologies are simple taxonomic structures of primitive or composite terms together with associated definitions. They are hardly axiomatized as the intended meaning of the terms used by the community is more or less known in advance by all members, and the ontology can be limited to those structural relationships among terms that are considered as relevant. This type of ontology are used like terminological resources in Natural Language Processing system, translation system or Information Retrieval system.

[3] http://www.culture.gouv.fr/documentation/thesarch/pres.htm

2.3 Classification according to Purpose

The classification is based on ontology uses: How the ontology is going to be used?

- Application Ontology: Used in a specific application implementing an inference engine based on an ontology. The typical trade-off between expressiveness and decidability requires a limited representation formalism. The Ontolingua server [6] or protégé server [9] are dedicated to the development of application ontology because they integrate reasoning capability to check the ontology.
- Reference Ontology: Used during development time of applications for mutual understanding and explanation between (human or artificial) agents belonging to different communities, for establishing consensus in a community that needs to adopt a new term or simply for explaining the meaning of a term to somebody new to the community. The PSL ontology which is a standard in building community is an example of reference ontology [12].

2.4 Classification according to Specificity

This set of ontology type define the role of the ontology to describe one or several domains.

- Generic Ontology: The concepts defined in this type of ontology are considered to be generic across many fields. Typically, generic ontologies (synonyms are "upper level" or "top-level" ontology) define concepts such as state, event, process, action, component etc. John F Sowa in his knowledge representation book define top level categories based on philosophical issues [16].
- Core Ontology: Core ontologies define concepts which are generic across a set of domains. Therefore, they are situated in between the two extremes of generic and domain ontologies. A core ontology of urban domain will contain generic concepts applicable to several urban field like roadway system or urban renewal. These generic concepts are means of transport, construction, building, material, urban form, politic action.
- Domain Ontology: Domain ontologies express conceptualizations that are specific for a specific universe of discourse. The concepts in domain ontologies are often defined as specializations of generic concepts of the related core ontologies. For example a domain ontology about road system contains concepts such as bus, traffic light coordinator, horizontal sign, coating courses. These precise concepts do not appear in other domain ontology.

3 Light Ontology: Towntology Project

The TOWNTOLOGY project[4] was initiated in 2002 between two laboratories at INSA of Lyon, one in computing (LIRIS) and one in urban planning (EDU). The EDU laboratory was in charge of developing and populating the ontology, whereas

[4] The first draft of the towntology project was financing by a BQR of the INSA of Lyon.

the computing laboratory was in charge of defining the structure and designing all software modules [14].

Towntology software allows the construction and the visualisation of a semantic network[5] of concepts. The concepts are in development, thus several natural language definitions or pictorial illustrations can be associated to a unique concept in order to express all possible interpretations. Ontologies[6] constructed by this tool are classified in lightweight ontology because no reasoning process is actually associated. Thus their expressiveness are limited.

Towntology ontologies are composed of concepts and relation types taxonomies. On top of textual definitions, concepts are explained towards relationship between them. If relation types are well defined and rigorously used the ontology formalization can be semi formal. On the other hands, if relation types have ambiguous or large definitions the ontology formalization is semi informal. For example, the "hierarchical" relation used in thesaurus has several meanings: it can be

- a "composition" relation linking a whole with its parts,
- a "specialization" relation linking a general topic with more specific ones.

The towntology browser offers three type of access to a concept [7]:

1. Select a term, which is a concept label in a alphabetical ordered list (cf fig. 2)
2. Navigate in a graph representing the semantic network in order to find the appropriate concept thanks to its relations with other concepts.(cf fig. 2)
3. Select part of images annotated by concepts. (cf fig. 4)
4. All the ontology are stored in an XML file.

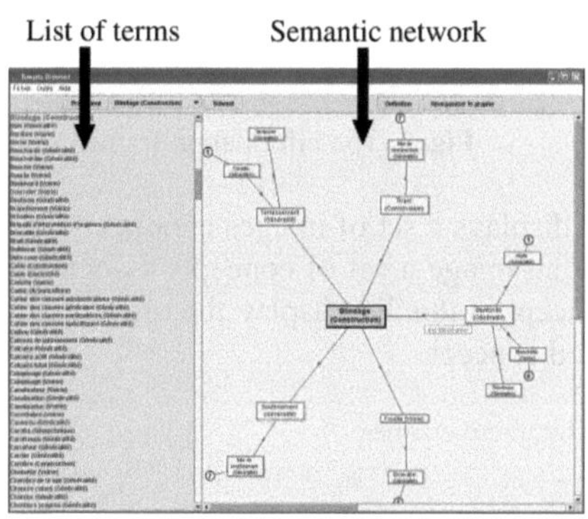

Fig. 2. The Towntology browser

The figure 2 presents the towntology browser, which interface is divided into two parts. The right part displays the semantic network. This graph is a part of the

---

5 A semantic network is often used as a form of knowledge representation. It is a directed graph consisting of vertices which represent concepts and edges which represent semantic relations between the concepts.

6 Actually ontologies constructing with towntology software relates to Gruber definition because we do not formalize.

semantic network of the whole ontology centered on a selected concept. This graph browser enables to navigate in the ontology. The left part is a scrollable list showing all the terms used in the graph as concept label. A click on a concept node and another click on definition button display the information frame (cf fig. 3) containing all the information about the concept.

The figure 3 shows the information frame displaying all information concerning public space concept. This concept has two definitions and is illustrated by an image. The source of definitions and images are also precised. On the left part of the information frame the list of relations having this concept as argument are listed.

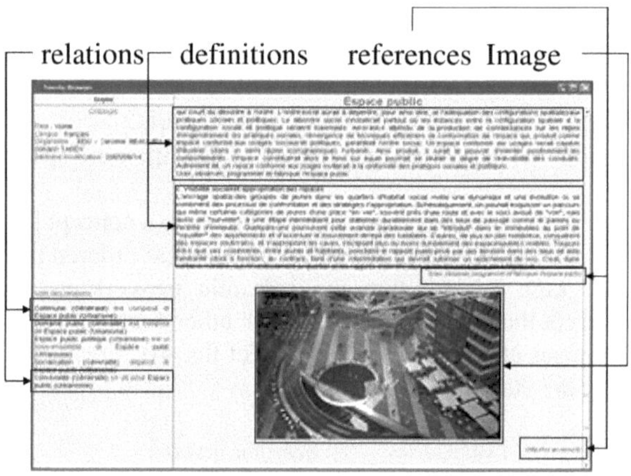

Fig. 3. The information frame

The image browser displays a set of images annotated by concepts. By clicking on the interactive zone of an image a set of concepts associated to this part of image is displayed. Select a concept in this list display the towntology browser with the graph centered on this selected concept.

Fig. 4. The image browser

Towntology software is a first draft to help user to build semi informal or semi formal ontology. Because no reasoning process is developed in the software the ontology expressiveness is limited as lightweight ontology. This software is still in development in order to integrate new functionalities based on user needs.

4 Cases Studies: 3 Examples of Urban Ontology Construction

The inexistence of ontologies for urban and town planning reinforces this project interest: that is to satisfy a need of coordination and cooperation between urban actors and software. Ontologies must limited semantics drifts of understanding. In fact, french words have the particularity to have different definitions dependent of the context. Thus, ontologies will take in account linguistic specificity to give to urban actors a consensual vocabulary. Forward, ontologies will improve dialogue between actors around a shared and formalized vocabulary. At last, operational urban ontologies will permit:

- improvement of urban services like road system management road system management or publics spaces management;
- synchronization and coordination of network intervention;
- consistent elaboration of different urban documents;
- impact definition of urban policies;
- …

Since 2002, 3 urban ontologies have been developed thanks to the collaboration of the EDU and LIRIS laboratories. They are designed for both no specialized and specialized users to clarify French urban domain. In consequence, the purpose of these ontologies is to become reference ontology. Moreover the topics of these ontologies is focused on specific fields of urban domain. So their specificity is to become domain ontology. The first ontology deals with a technical domain easily modelled. Moreover, the terminology was unequivocal and consensual. The construction of the second ontology has been more difficult because the domain was more fuzzy: the term interpretation was dependent of context. Thus, relationships between concepts were difficult to fix. In the third ontology, these difficulties were more and more important due to appearance of social aspect. Those aspects are more difficult to model. Moreover the causality relationship between different events could be difficult to establish. In consequence, concepts are more complex and relations are less precise.

4.1 Road System Ontology

1000 terms were defined in the road system ontology [1]. This vocabulary come from several technical dictionaries. The technical domain of roadway system have permitted to associate to each term one concept with a precise definition taken from referenced documents. From those concept definitions, 21 types of relationship have been identified to link concepts between them. These relations represent link expressed in term definitions. A subset of semantic network using 21 types of relationship is presented in Figure 5.

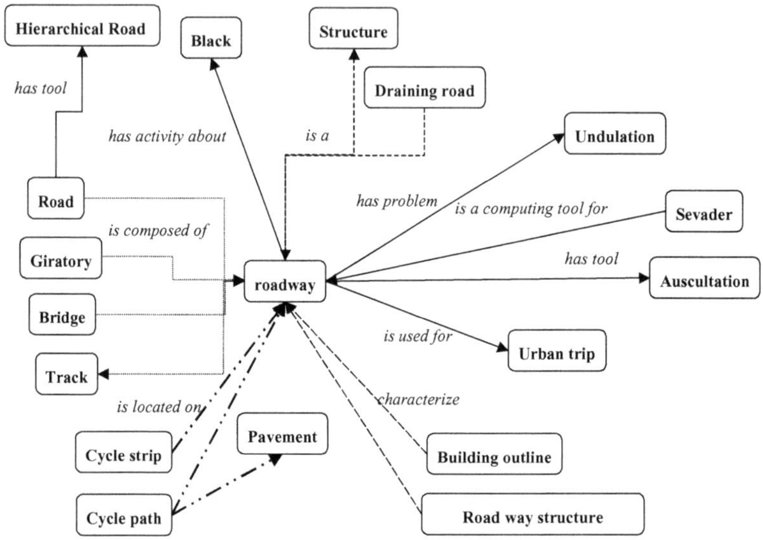

Fig. 5. Part of the road system ontology

The road system ontology is not only constituted with concept linked by characterizing relationship, but on top of the semantic network, pictures have been utilized to illustrate concepts. The following figure permits to visualize road system concepts in their context. Thanks to the image browser, this annotated image enabled to access to the associated semantic network.

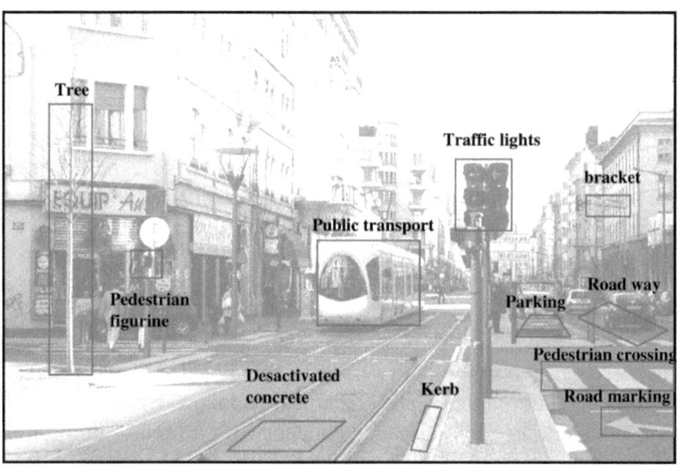

Fig. 6. An image annotated with road system concepts

The work of ontology development wasn't preceded by a reflection work about the nature of relationship we could establish between concepts. When this first ontology will be associate with another one, the number of relationship and their redundancy must become a problem. The multiplicity of relationship make difficult the graph visualization. This situation forces us to simplify the set of relationship types, which are restricted to 7 classes. As a consequence, the 21 types of relation were limited to 12 types of relations presented in the following table.

Table 1. The 7 classes of relationship types.

1	Relation of localization	« is located on »
		« is located in »
2	Relation of use	« is used for »
		« is used by »
		« can hold the role »
3	Relation of composition	« is composed of »
4	Relation of subordination	« depends on »
		« works for »
5	Relation of being	« is a(n) »
6	Relation of characterization	« is characterized by »
		« says itself for »
7	Relation of generation	« is resulting from »

This first ontology development experimentation shows some limits. The first problem we encountered during the ontology development process is the cohabitation of different level of vocabulary (expert, beginner, ...). The question is how to manage synonym to keep vocabulary coherence? Could urban ontologies represent the complexity of urban domain? The road system ontology has given a partial response: Ontologies have shown their interest and their feasibility.

4.2 Urban Mobility Ontology

The second experience of ontology construction aims at integrate abstract concepts of mobility and trip [3]. The objectives were to verify the capacity of ontologies to represent general data liable to various interpretation. In this ontology 200 concepts have been defined: 190 concepts concern mobility and trip and 10 concepts are used like foot bridge toward road system ontology. These new concepts are found by inquiring several types of mobility experts like designers, mobility network designers, city council and so one. Firstly, each of inquired people had to give a list of core concepts about mobility. Secondly, the ontology designer associated to each concept several definitions found in reference documents like dictionaries.

Afterwards, inquired experts had to validate or complete definitions. Finally they had to give their opinions about the concept network build. The semantic network has been represented like that.

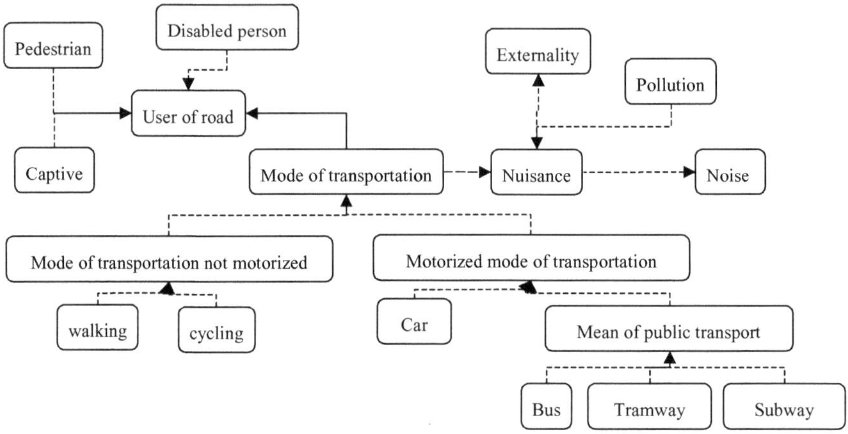

Fig. 7. Part of the mobility and trip ontology

Examples of foot bridge concepts are noise, trip, urban, road, parking. Those concepts are used to link two ontologies. Thus they belong to both ontologies.

The second ontology construction experimentation has shown several difficulties: the two first ones concern concept definition and the third one is about the creation of semantic network.

1. Some concepts relating to mobility and trip haven't official definition because consensus is in construction. For example, the French concept of «mode concourrant » has not a fixed definition, due to the fact that some research about this subject is still in development. We notice that concepts without consensual definition make problems: Their definitions are ambiguous and their interpretations are dependent of the context; Moreover, relations between these concepts are also dependent to the technical sensibility of the author.
2. Difficulty to choose metadata to characterize definitions. Those metadata describe the domain activity of the concept (for example computing or legal domain) and the level of definition specialization (for example expert or beginner).
3. Difficulty to choose relevant and no ambiguous set of relationships. For example, too much specific relationship or too much personal relationship do not permit to create a consensus. In this case the ontology cannot be used by other people of the domain.

The ontology construction with no consensual concepts is possible when we make reasonable and targeted choice. Of course, this type of ontology represent a simplified vision of the domain, but it could be useful for a first approach.

Face to domain in development, it is essential to identify invariant concepts (where a consensus exist) from concepts in development. It is necessary to identify these concept in development in order to follow up their update.

4.3 Urban Renewal Ontology

Urban renewal ontology is in construction [2]. Actually, she has only 30 concepts. Like the mobility ontology, our aim is to incorporate abstract and evolving concepts. In a first step, this ontology describes only the French case, but afterwards other cases can be incorporated. In France, urban renewal mix up with high rise estate. Urban renewal is an operation starting with demolition in order to rebuild. This ontology is the representation of a specific point of view of an expert in urban renewal. Thus only one designer has to build the concept network.

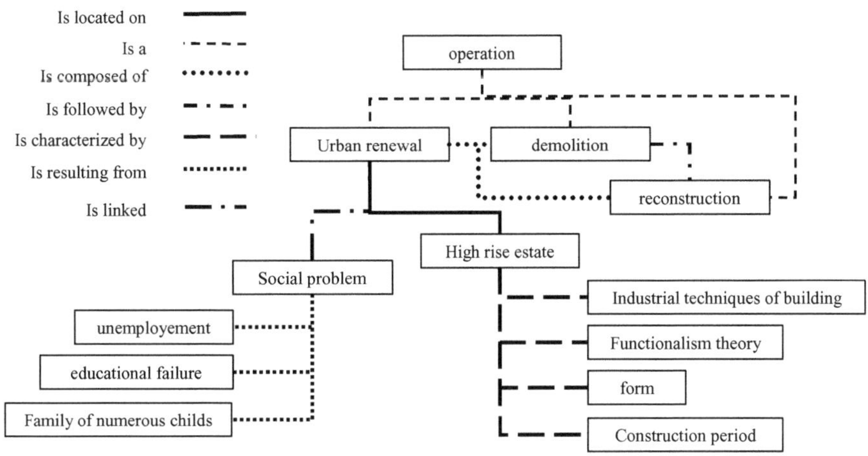

Fig. 8. The Urban Renewal Ontology

This ontology describes a conception of urban renewal more ambitious than local «social development» policies. It would like to cover every area of urban development, and all policies.

Urban renewal will integrate three dimensions:

- The disciplines involved (economics, social affairs, culture, physical development, ecology, etc.);
- The geographical and institutional areas of definition and implementation;
- The various time scales (for the resident, for the investor, for the politician, etc).

The urban renewal ontology integrates several aspect:

- an juridical approach
- an historical approach
- an economical approach
- an geographical approach
- an institutional approach
- …

At least it would be possible to compare different urban renewal policies in Europe with an historical, economical, aspect.

5 Conclusion

Ontologies has the same drawback than mathematic models. Their objectives are to produce an accurate representation of the reality but in order to reach this goal they must simplify this reality. This constatation being made, the choice of modelling urban domain rises some questions.

If we conclude that technical domain are complicated to model, what will be the conclusion for complex domain?

The construction of the 3 ontologies have been an opportunity to make choice and to query ourselves about urban concepts. Our questioning were about the definition of these concepts and the nature of relationship between them. This research interest have forced us to take in account our researcher subjectivity and have reinforce our interest to validate our choice by a working group.

However, the ontology experiment has revealed its flexibility and permits unexpected usage such as book cartography for library, chronological classification, genealogical classification. Ontologies are also very useful for the comprehension of urban law (SRU law in France). Moreover, evaluations of the usability of the software and associated ontologies are positive [15]. The market about this tool box is not yet exploited, but user needs have to be clarified.

This software is now a very pertinent information tool and can be an assistant tool for decision making on targeted public. Never the less, it is too limited to be sufficient for teaching urban logics.

References

1. Beaulieu, C., Tardy, C. Projet Towntology: construction d'une ontologie. PIRD GCU EDU. 48 p. 2003
2. Berdier C. Urban renewal: how to make a comparison between different approachs: case studies: France, Italy and Spain. Cost C21 meeting. Belfast 8 & 9 mai 2006.
3. Berthet, C. Projet Towntology: integration de concepts lies à la mobilité. PIRD GCU EDU. 68 pages. 2004
4. Crubézy, M. and Musen, M.A. (2003). Ontologies in Support of Problem Solving. Handbook on Ontologies. S. Staab and R. Studer, Springer: 321-341.
5. Dublin Core Initiative http://dublincore.org/
6. Farquhar A., Fikes R., Rice J. The Ontolingua Server: A Tool for Collaborative Ontology Construction. Knowledge Systems, AI Laboratory, 1996.
7. Keita A. K., Roussey C., Laurini R. Un outil d'aide à la construction d'ontologies pré-consensuelles: le projet Towntology à paraître dans les proceeding inforsid 2006.
8. Knowledge Interchange FormatVersion 3.0 Reference Manual M. R. Genesereth, R. E. Fikes (Editors). Computer Science Department, Stanford University, Technical Report Logic-92-1, June 1992.

9. Knublauch H. Ontology-Driven Software Development in the Context of the Semantic Web: An Example Scenario with Protégé/OWL. International Workshop on the Model-Driven Semantic Web, Monterey, CA, 2004.
10.Gruber T. R. A translation approach to portable ontologies. Knowledge Acquisition, 5(2):199-220, 1993
11.http://cos.ontoware.org/
12.Pouchard, L., Cutting-Decelle, A.F., Michel, J., Young, R., Das, B., "Utilizing Standard-based Approaches for Information Sharing and Interoperability in Manufacturing Decision Support," Proceedings of the Flexible Automation and Intelligent Manufacturing Conference, Toronto, Canada, July 12-14, 2004.
13.Resource Description Framework is a lightweight ontology system to support the exchange of knowledge on the Web http://www.w3.org/RDF/
14.Roussey C., Laurini R., Beaulieu C., Tardy Y., Zimmerman M. Le projet Towntology: un retour d'expérience pour la construction d'une ontologie urbaine. Revue Internationale de Géomatique 14(2):217-237, 2004.
15 Simonnot, N., Rascol R. Projet Towntologie: complément d'intégration de concept liés à la mobilité et évaluation auprès d'utilisateurs. PIRD GCU EDU. 52 p. 2005
16.Sowa, J. "Top-Level Ontological Categories". In International Journal on Human-Computer Studies, 1995, Vol. 43, N° 5/6, pp. 669-685.
17.Uschold, M. and Gruninger, M. Ontologies: Principles, Methods and Applications. Knowledge Engineering Review 11(2) 1996.

Building an Address Gazetteer on top of an Urban Network Ontology

J. Nogueras-Iso, F. J. López, J. Lacasta, F. J. Zarazaga-Soria, and P. R. Muro-Medrano

University of Zaragoza, Zaragoza, Spain
{jnog,fjlopez,jlacasta,javy,prmuro}@unizar.es

Abstract. In order to create the contents of an address gazetteer service that forms part of a city council Spatial Data Infrastructure, all the existent repositories containing address information in the different council offices must be analyzed and harmonized. The problem is that usually these repositories are constrained by the use of different taxonomies for the identification of urban network feature types. The objective of this work will be to describe how to establish a formal ontology enabling the interoperability among the different taxonomies, and facilitating the construction of the gazetteer contents.

1 Introduction

The increasing relevance of geographic information for decision-making and resource management in diverse areas of government has promoted the creation of Spatial Data Infrastructures (SDI), which are usually defined as a coordinated approach to technology, policies, standards, and human resources necessary for the effective acquisition, management, distribution and utilization of geographic information at different organization levels and involving both public and private institutions. In the particular context of the development of an SDI for local administrations such as a city council, address gazetteer services represent one of the most important services that the councils must offer to their citizens [1]. The councils are responsible for the management of urban networks, and these networks are used as reference information for other services at national level such as cadaster or census services.

The creation of contents for an address gazetteer service requires SDI developers to perform a work of analyzing and harmonizing all the existent repositories containing address information in the different offices of the council. The main problem typically found is that different taxonomies are used for the identification of urban network feature types in different administrative processes. Frequently, when city councils need to exchange information with external organizations like National Cadaster Offices or National Statistics Institutes, the information needs to be reformatted in order to comply with the feature types accepted by these institutions. Moreover, it is usual that this reformatted information is stored at council level in parallel repositories (e.g., tax office databases, urban planning office databases) whose updates are not synchronized.

J. Nogueras-Iso et al.: *Building an Address Gazetteer on top of an Urban Network Ontology,* Studies in Computational Intelligence (SCI) **61**, 157-167 (2007)
www.springerlink.com

In order to overcome the existent heterogeneity in the different repositories used for gazetteer contents, it seems sensible to establish a unified model of the feature types that can be found in this domain, and make the necessary mappings to the particular taxonomies that must be used in external organizations or in the different repositories maintained at council level. This feature type model could be formally represented by an ontology that defines explicitly the concepts and relationships between these concepts in a domain [2, 3]. On the one hand, this unified ontology would facilitate the interoperability with external administrative organizations. And on the other hand, it would enable the modelling of the contents served by the Gazetteer service.

Having observed this necessity of defining an ontology for feature types in the urban networks domain, the objective of this work will be to explore the mechanisms to build a unified urban network ontology on top of the existent taxonomies in the public administration for urban networks. The construction of an ontology upon existing vocabularies (textual dictionaries; glossaries; or even more structured vocabularies that can include taxonomies, thesauri or other existent ontologies) is a classical and widely used approach in ontological engineering whose main problem is how to reconcile the source taxonomies. For instance, [4] proposes the creation of a unified ontology by mapping different source ontologies against a common controlled vocabulary (the ADL Feature Type Thesaurus). The underlying problem, also known as *ontology alignment*, is how to find the relationships (e.g., equivalence or subsumption) that hold between the entities represented in different taxonomies. Moreover, ontology alignment methods may be useful for assisting conflict resolution among people having different conceptualisations of a given domain.

The remaining paper is organized as follows. Section 2 analyzes the use-case selected for this work explaining the different urban network databases (including their different feature type taxonomies) that must be used for the creation of a gazetteer. Then, the next two sections describe how to build the ontology that will guide the contents of the gazetteer. Whereas the first approach will describe an ad-hoc manual mapping among taxonomies used in the source repositories, the second one will describe how to apply *Formal Concept Analysis* techniques for the automatic creation of a formal urban network ontology that integrates the mappings among the different taxonomies. Finally, the paper ends with some conclusions and future lines.

2 Use case: Urban Network databases at the Zaragoza city council

The use-case selected for this work has been the SDI developed for the Zaragoza city council in Spain (IDEZAR, http://www.zaragoza.es/idezar/). Figure 1 sketches the flow of information concerned with urban transport networks between the different offices of the Zaragoza city council and from/to external administrative bodies. Three different categorizations are used for the urban network feature types in the different offices of the Zaragoza city council: *SIGLA*, *TVIAN* and *AYTO*.

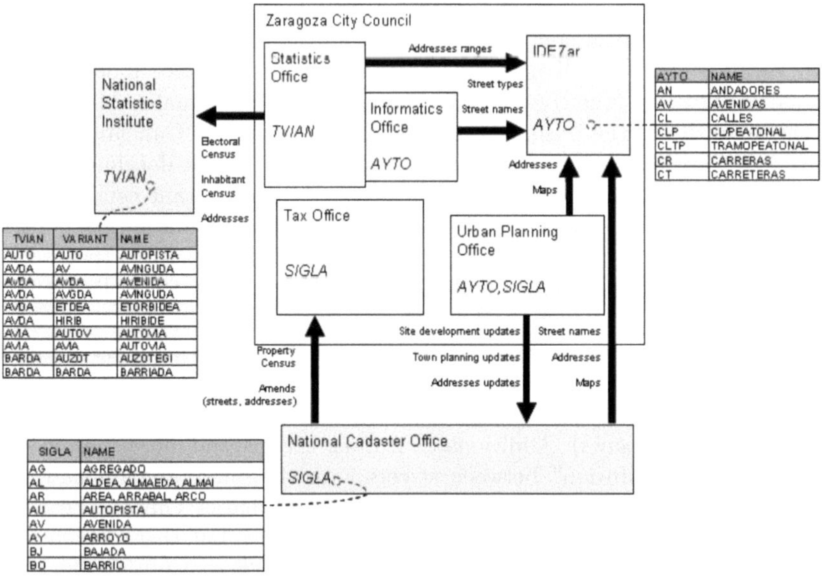

Fig. 1. Workflow

The *SIGLA* categorization (*Sigla de vía pública*) is a code list that describes the street types used in the transfer file format between the local government and the National Cadaster Office. This code list consists of acronyms and some of them are shared by two different street types (e.g., *CM* represents both *Camino/path* and *Cármen/southern kind of house*). *SIGLA* is used in the urban network databases managed by the Tax Office and the Urban Planning Office of the Council. Whereas the Tax Office is responsible of land taxes management, the Urban Planning Office is responsible of the land development. Its Geographical Information Service is responsible for the urban cartography and the parcel numbering. As regards the data flows where *SIGLA* is involved, three main data flows can be mentioned. Firstly, *SIGLA* is used for land address oriented data flows. The Tax Office informs the National Cadaster Office about tax management (owner addresses) and land management (property addresses). Secondly, *SIGLA* is used for tax management data flow. This flow comprises all the data sent from the National Cadaster Office to the city councils to help land tax management (property taxes, land valuation changes, amends). Also city councils may inform the National Cadaster office any change in the owner's data or mistakes. And thirdly, *SIGLA* is also used for town planning and development data flow. This flow informs the National Cadaster Office about any land related permissions, planning change or address change made by the city councils.

The second categorization, *TVIAN* (*Tipo de Vía Normalizado*) is a partially normalized code list of street types used in the transfer file format between local governments and the Spanish National Statistics Institute. It establishes a mapping between a normalized key and a set of acronyms, which are variants in the different local languages. However, there is no hint of the language of each variant and, it is ill normalized as some concepts have more than one normalized key (e.g., the concept *callejón/alley* has the normalized key *CLLON* for the Spanish and the Basque language but *CXON* for the Galician language). *TVIAN* categorization is internally used in the council for the database managed by the Statistics Office, which is responsible of the inhabitant census and the poll census. As regards data flow, *TVIAN* is involved in the data flow concerned with citizen statistics. This flow, which goes from city councils to the National Statistics Office, comprises the inhabitant continuous census, the poll census and any change in streets, street number ranges and addresses.

And the third categorization is called *AYTO*. This code list is owned by the Zaragoza council which compiles the street types included in the local regulation (e.g., *caminos/paths*, *carreteras/roads*, *plazas/squares*, *calles/streets*, *paseos/boulevars*, *parques/parks*). Additionally, it integrates as well more specific street types "to avoid confusion" between streets with the same name. This categorization is used by the Culture Office and the Informatics Office. The council street names are proposed by the Culture Department, but their encoding and maintenance is the responsibility of the Informatics Office. Additionally, the Informatics office gives technical support to the applications based on the council gazetteer in hardcopy version.

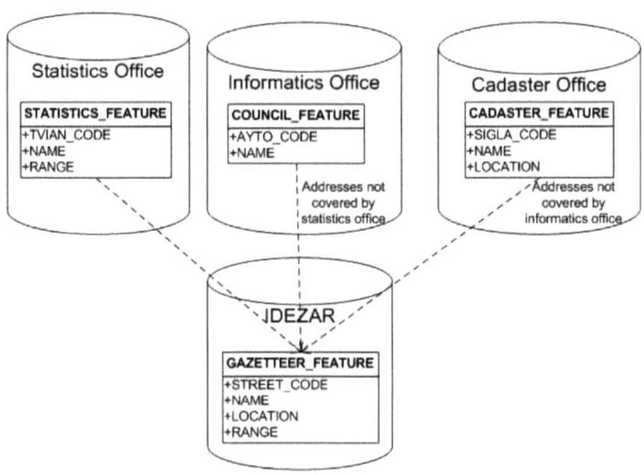

Fig. 2. Gazetteer contents

Finally, extracting the information from the three databases described above, the objective was to define an electronic gazetteer aggregating the information available in such databases. Figure 2 shows how the conceptual models from the above databases are matched and merged for the creation of the conceptual model that would define the contents of the desired electronic gazetteer. However, the main problem remains at the instance level in order to match the feature instances from the source databases and produce the feature instances the must be uploaded in the gazetteer database. That is to say, for each feature instance (e.g., *Plaza España*) found in both source databases, we need to integrate the normalized street name (from *COUNCIL_FEATURE* database at the Informatics Office) with the location (from *CADASTER_FEATURE* database at Tax Office) and the street range (*tramero* found in the *STATISTICS_FEATURE* database at the Statistics office). And for that purpose, it seems obvious that we need a mapping between the different feature types found in the three categorizations. On one hand, the matching of feature instances is not enough just using the feature names. For instance, *Espana* is a name that can be used for squares and streets. And on the other hand, it would be interesting that the gazetteer provides a consistent feature type categorization for the features served by the gazetteer, probably the common factor of the three source categorizations.

3 Ontology construction using a manual mapping approach

As explained in the introduction, we must face the problem of aligning the different taxonomies already available in order to identify equivalences between the entities represented in the different taxonomies and extract the most relevant concepts (including as well possible subsumption hierarchies).

One approach for this alignment is obviously the manual mapping between the different taxonomies. In particular, given that the taxonomies mentioned in previous section (*TVIAN*, *AYTO*, *SIGLA*) had no semantic description (in most cases just an acronym and the complete name), a manual mapping approach was tried in first place. That is to say, a human expert had the responsibility of comparing terms (acronyms+names) in the different taxonomies and establishing the mapping across the different taxonomies.

The objective was to use the *AYTO* as the bridge between the taxonomies and establish the manual mappings: *SIGLA* - *AYTO* (see figure 3), and *TVIAN*-*AYTO*. As *SIGLA* belongs to a property-oriented database, *TVIAN* belongs to a census-oriented database and *AYTO* belongs to an urban oriented database, it was expected that identical terms would overlap in the different databases. However, even in this specific domain, it was found that homonyms can arise (even with terms belonging to the same conceptual design).

The procedure for the manual mapping consisted of the following steps: collect acronyms from the different database; expand acronyms with their complete names; look up for definitions; and match equivalent terms based on their similar definitions. The matches that were obtained could be classified in the following categories:

- Exact match: the meanings of both concepts are identical.

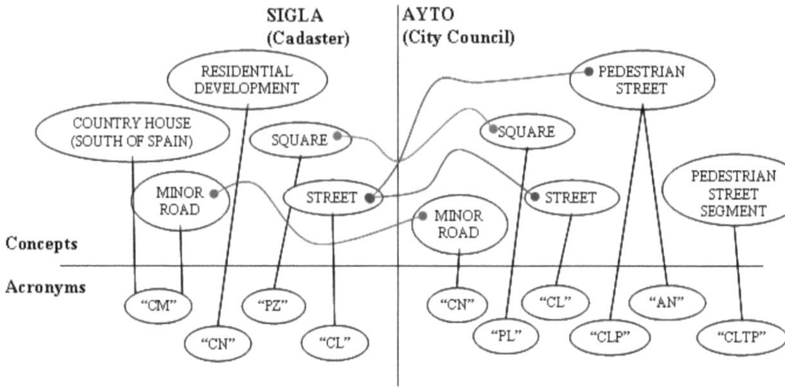

Fig. 3. Mapping *SIGLA-AYTO*

- Partial match: one concept is broader or narrower than the other. The concept represented by *CL* (street) in *SIGLA* has narrower concepts in *AYTO* such as the concepts represented by *CLP* and *AN* (different types of pedestrian streets).
- Provisional match: due to the design of *SIGLA* where different concepts share the same acronyms, the matching of concepts is provisional. The feature instances linked with this match should be verified.
- No match: the concept does not exist. We need add a residual category to cover these cases.

The experience from this first approach has shown that this non-systematic manual process results quite subjective, too time expensive and with little scalability. If a new taxonomy is added to the possible lists, a new mapping to the not very well structured *AYTO* taxonomy should be established. A more flexible approach could be the use of well-established shared common core and make mappings between the distinct sources and this common core.

Thus, our second experiment consisted in mapping the source taxonomies against the URBISOC thesaurus [5]. It is a thesaurus focused on Spanish terminology for Town Planning, which has been developed by the CINDOC/CSIC institute (Centre for Scientific Information and Documentation/Spanish National Research Council) to facilitate classification at the URBISOC bibliographic database, which is specialized in scientific and technical journals on Geography, Town Planning, Urbanism and Architecture. Additionally, we decided to use a proper ontology editor to facilitate communication and discussion between the experts in charge of the alignment. The tool selected was Towntology [6], which enables the storage of the ontology, the display of the ontology in visual graphic form, to navigate in the ontology and to query it. The main difference with respect to other ontology editors such as Protégé [7] is that it is

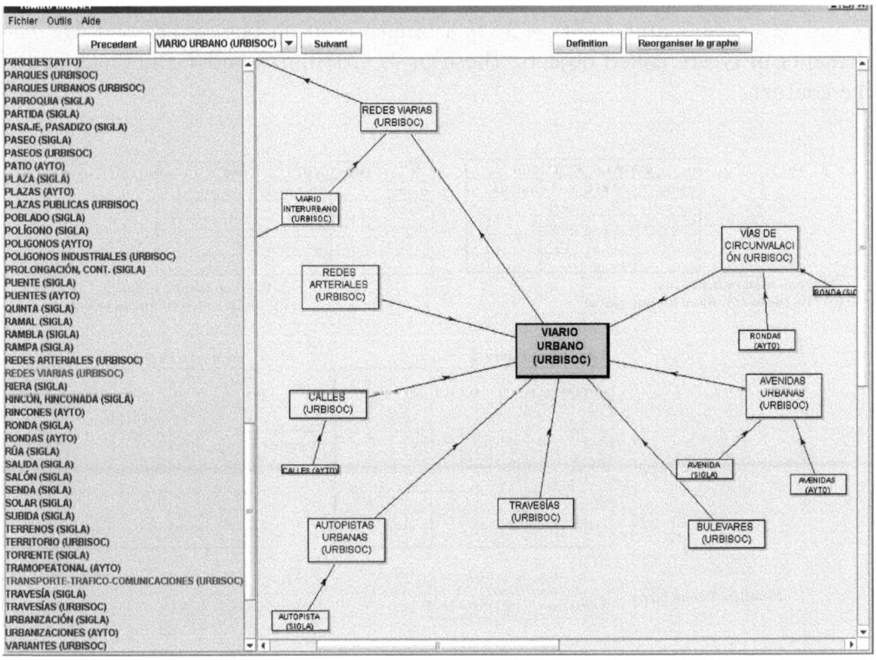

Fig. 4. Use of URBISOC as common core

not based on any formalism such as RDF(s) [8] or OWL [9]. But at this step we were more focused in the ontology construction process than in representing formally a built ontology. The Towntology tool is aimed for storing concepts with several definitions that are in a process of selection and characterization of these definitions.

Figure 4 shows a screenshot of the Towntology browser displaying some mappings between the URBISOC thesaurus and some of the terms available at *SIGLA* and *AYTO*. Although improving the scalability, this second attempt results still time expensive and error prone.

4 Ontology construction using an FCA approach

Having seen the difficulties in establishing a manual mapping of ontologies, it is highly beneficial to count on methods for the automatic alignment of these existent vocabularies, facilitating the rapid creation of a draft of the desired ontology.

In particular, this section describes the applicability of *Formal Concept Analysis* (FCA) techniques [10, 11] to output a hierarchy of concepts from the feature instances contained in the three databases shown in section 2. The basis of FCA is the definition of a *formal context*, which consists in a triple (G, M, I) where

G and M are sets and $I \subseteq G \times M$ is a binary relation between G and M. The elements in G are called objects, those in M attributes and I the incidence of the context.

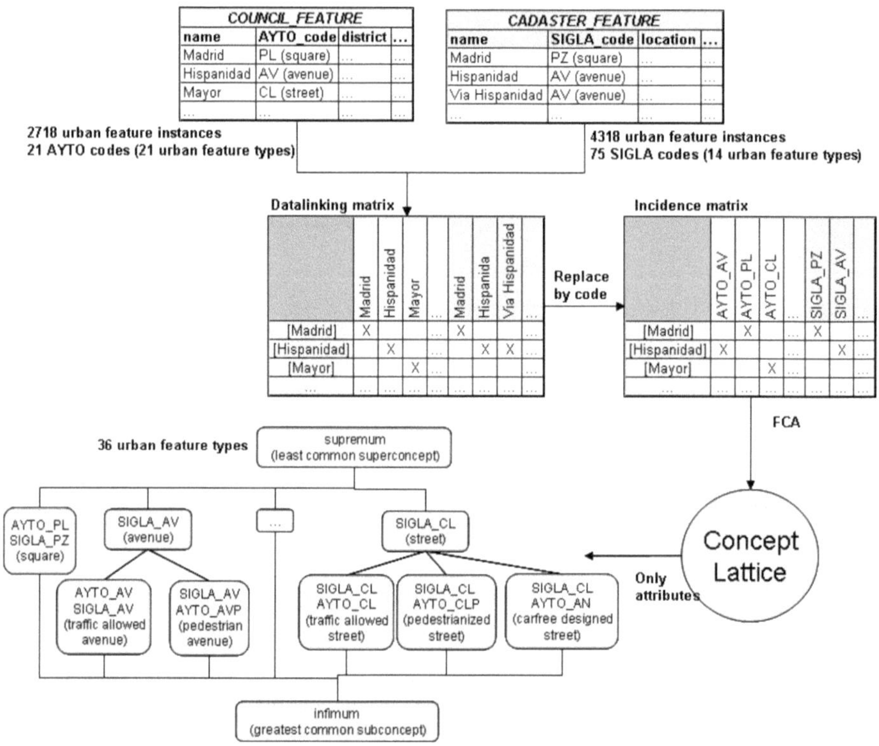

Fig. 5. Application of FCA

The objective of FCA is the extraction of a lattice of formal concepts, but previous to the definition of formal concepts we will define A' and B' for $A \subseteq G$ and $B \subseteq M$:

$$A' = \{m \in M | (g, m) \in I \ for \ all \ g \in A\}; \ B' = \{g \in G | (g, m) \in I \ for \ all \ m \in B\} \quad (1)$$

A' can be understood as the set of all the attributes common to the objects in A and B' is the set of all the objects which have in common with each other the attributes in B. And given these definitions, a pair (A, B) is a formal concept if and only if

$$A \subseteq G, B \subseteq M, A' = B \wedge A = B' \quad (2)$$

In other words, (A, B) is a formal concept if and only if the set of all attributes shared by the objects in A is identical with B and on the other hand A is also the set of all the objects which have in common with each other the attributes in B. Furthermore, the concepts of a given context are naturally ordered by the subconcept-superconcept relation defined by

$$(A_1, B_1) \leq (A_2, B_2) \Longleftrightarrow A_1 \subseteq A_2 \; (\Longleftrightarrow B_2 \subseteq B_1) \tag{3}$$

The ordered set of all formal concepts of (G, M, I) enables the definition of a concept lattice by linking it is possible to establish a concept lattice.

Figure 5 depicts the process of applying the FCA techniques to the instances contained in two source databases: the *COUNCIL_FEATURE* database using *AYTO* taxonomy, and the *CADASTER_FEATURE* database using *SIGLA* taxonomy. The main problem for the direct application of FCA techniques in our context was how to obtain a unique repository of instances, i.e. the formal context required by FCA. Therefore, in order to obtain this unique repository, traditional datalinking has been applied to the feature instances contained in the different databases. This datalinking has been based on the analysis of the lexical and spatial similarities of feature attributes, i.e. the lexical similarity of names (use of *SecondString* string similarity function library [12]) and the proximity of locations. Then, the datalinking matrix obtained as a result of this process together with the transformation of urban network feature type codes (e.g., *AYTO_CODE*, *SIGLA CODE*) into proper attributes (with boolean values) enables the creation of the incidence matrix I of the formal context.

Once obtained the incidence matrix, a version of the algorithm *next closed set* [13] has been used to generate the concept lattice that establishes the alignment between the two source taxonomies. Thanks to the FCA technique and some minor adjustments, the source taxonomies can be transformed into a merged hierarchy of formal concepts. The technique not only identifies equivalent concepts in both taxonomies, but also subconcept-superconcept relations. An example of equivalent concept would be a *square* (*PL* in *AYTO* and *PZ* in *SIGLA*). And an example of subconcept-superconcept relation would be the identification of *street* as a broader concept in *SIGLA* (*CL*), which has narrower concepts in the *AYTO* taxonomy such as traffic-allowed streets (*CL*), pedestrianized streets (*CLP*) or carfree-designed streets (*AN*).

5 Conclusions

This paper has shown different mechanisms for the construction of an urban network ontology by means of the alignment of different source taxonomies. In particular, a manual mapping approach and an automated approach based on *Formal Concept Analysis* have been studied. Although minor problems must be supervised manually, it has been demonstrated that the second approach (based on FCA) provides more flexibility and scalability. Additionally, this technique enables the extraction of concepts independently from the encoding of the feature types. That is to say, it would be possible to analyze different data sources that have used a number encoding, without any apparent meaning, for the classification of features.

Besides, the unified ontology obtained as result of this alignment process has been used to create the contents of an address gazetteer service integrated within the SDI of a local council. The unified ontology enables the union of the toponyms coming from the different databases used in the city council offices, detecting when necessary the intersections and avoiding duplications. Furthermore, this unified ontology would allow the construction of customized user query interfaces which can still use the original taxonomies according to the requirements of each city council office.

As future lines of this work, it is planned to make a refinement of the FCA-based approach in order to improve the efficiency and the formalization of the generated ontologies. On the one hand, it is believed that the detection and filtering of instances that may introduce noise will avoid generating spurious concepts. On the other hand, the formalization level could be enriched by means of extracting statistics over the attributes of original feature instances (e.g., conclusions about limits on the perimeter, area or geometry of the *square* concept). Finally, it is worth noting that this FCA-based approach could be also applied to other domains making use of toponyms and where ontologies help revealing the structure of separate repositories. For instance, it could be applied to the analysis of hydronyms, which are usually managed at national and regional levels by National Mapping Agencies and Water Agencies respectively.

Acknowledgements

This work has been partially supported by the Spanish Ministry of Education and Science through the project TIN2006-00779 from "the National Plan for Scientific Research, Development and Technology Innovation" and by the COST UCE C21 Action (Urban Ontologies for an improved communication in Urban Civil Engineering projects) of the European Science Foundation.

References

1. Portolés-Rodríguez, D., Álvarez, P., Muro-Medrano, P.: IDEZar: an example of user needs, technological aspects and the institutional framework of a local SDI. In: Proc. 11th EC GI & GIS Workshop, ESDI Setting the Framework. (2005)
2. Gómez-Pérez, A., Fernández-López, M., Corcho, O.: Ontological Engineering. Springer-Verlag, London (United Kingdom) (2003)
3. Visser, P.R.S., Jones, D.M., Bench-Capon, T.J.M., Shave, M.J.R.: An Analysis of Ontological Mismatches: Heterogeneity versus Interoperability. In: AAAI 1997 Spring Symposium on Ontological Engineering, Stanford, USA (1997)
4. Berman, M.L.: Semantic Interoperability and Cultural Specificity: Examples from Chinese, Japanese, Mongolian and Uighur. In: Proc. of Social Science History Association meeting (SSHA'2003), Baltimore (2003)
5. Alvaro-Bermejo, C.: Elaboración del Tesauro de Urbanismo URBISOC. Una Cooperación Multilateral. In: Encuentro Hispano-Luso de Información Científica y Técnica. II, Salamanca (1988)

6. Keita, A., Laurini, R., Roussey, C., Zimmerman, M.: Towards an Ontology for Urban Planning: The Towntology Project. In: CD-ROM Proc. 24th UDMS Symposium, Chioggia (2004)
7. Noy, N.F., Fergerson, R.W., Musen, M.A.: The knowledge model of Protege-2000: Combining interoperability and flexibility. In: Knowledge Engineering and Knowledge Management. Methods, Models, and Tools: 12th Int. Conf., EKAW 2000. Volume 1937 of LNCS., Juan-les-Pins, France (2000) 17–32
8. Manola, F., Miller, E., eds.: RDF Primer. W3C, W3C Recommendation 10 February 2004 (2004) http://www.w3.org/TR/2004/REC-rdf-primer-20040210/.
9. Bechhofer, S., van Harmelen, F., Hendler, J., Horrocks, I., McGuinness, D.L., Patel-Schneider, P.F., Stein, L.A.: OWL Web Ontology Language Reference. W3C, W3C Recommendation 10 February 2004 (2004) http://www.w3.org/TR/2004/REC-owl-ref-20040210/.
10. Ganter, B., Wille, R.: Formal Concept Analysis: Mathematical Foundations. Springer, Berlin-Heidelberg (1999)
11. Stumme, G., Maedche, A.: FCA-MERGE: Bottom-up merging of ontologies. In: Proc. 17th IJCAI, Seattle (WA US) (2001) 225–230
12. Cohen, W.W., Ravikumar, P., Fienberg, S.E.: A Comparison of String Distance Metrics for Name-Matching Tasks. In: Proc. IIWeb 2003 (IJCAI 2003 Workshop). (2003) 73–78
13. Ganter, B.: Algorithmen zur formalen begriffsanalyse. Beiträge zur Begriffsanalyse. BIWissenschaftsverlag, Mannheim/Wien/Zürich (1987) 241–254

"Pre-Ontology" Considerations for Communication in Construction

John Lee[1,2] and Dermott McMeel[1]

[1]University of Edinburgh, School of Arts, Culture and Environment Graduate School, Alison House, Nicolson Square, Edinburgh, UK
[2]Human Communication Research Centre, 2 Buccleuch Place, Edinburgh, UK
J.Lee@ed.ac.uk, D.McMeel@sms.ed.ac.uk

Abstract. We address the question of how the use of ontologies can assist communication in construction. We consider the background to this question in contemporary and emerging practices, and contend that due to the particularity of building site contexts current approaches involving broad standardisation are not promising. We argue that ontologies should be thought of as higher-level conceptual tools for revealing areas of disagreement or differences in understanding. We propose that their use could usefully be integrated with a system for capturing negotiation aimed at resolving these differences.

Keywords: Communication, construction, ontology, negotiation

1 Introduction

In this paper, we address the question of how the use of ontologies can assist communication in construction. We consider the background to this question in contemporary and emerging practices, both on building sites and among those working to develop assistive technologies. We observe that, to the extent these practices currently meet, the meeting is not as fruitful as it might be, and we seek to identify possible ways to improve it. Refined uses of ontologies are often thought to be a key factor in achieving such improvement; we examine this idea and conclude that some promising uses of ontologies might be rather different to those usually envisaged.

2 The Building Site

Our focus in this paper is on the building site, especially where a large project is in progress. This focus is natural given our interest in communication. On a large building site, there are many activities that bring out and emphasise issues of communication. The overall pattern is very fluid; little stays the same throughout the

J. Lee and D. McMeel: *"Pre-Ontology" Considerations for Communication in Construction,* Studies in Computational Intelligence (SCI) **61**, 169–179 (2007)
www.springerlink.com

construction project, and few people may be involved in the whole process from beginning to end. For example, there is commonly a succession of specialised contractors who arrive on the site to carry out some particular task (which may take hours, days or weeks), and then leave. While present, they need to interact with those already there, and with others who may arrive. They may bring their own specialised workers and equipment, but also may use equipment otherwise in place (tower cranes, perhaps). They will have to interact with the site and construction state as they find it, and will probably change it (by carrying out earthworks or installing services) or add to it (by erecting some component, installing a cladding system, etc.). Often, they will bring and install building elements that have been manufactured off-site, and this may well involve finding that there are discrepancies in detail between their expectations and what is actually present on the site; there may need to be adjustments and alterations.

Whatever happens, there will almost inevitably need to be discussion and negotiation. And this discussion or negotiation has a number of interesting characteristics, such as the following.

1. It takes place very often between strangers. Contractors tend to move from site to site, carrying out broadly similar operations for different projects. Few of the contractors involved in a project may have worked together before. The nature of the project, and its similarity with other projects, will of course provide a shared context for any discussion, but still those involved need initially to some extent to feel their way towards an understanding of the other's apprehension of any given problem.
2. It is local and specific to the situation on a given site. No matter how similar two projects may be, what happens on a site is always affected by the detail of the site itself, the particular nature of the things that have been assembled on it, and the individuality of those who have been involved.
3. It occurs in the context of a dense web of regulations, documents, standards, procedures, and against a background of increasingly litigious demands for performance to contract. The construction industry is very richly provided with documentation at all stages of a project, and the framework of laws and regulations surrounding site practices, design requirements, materials specifications, and almost every other aspect of the activity, is enormous. Very little is supposed to happen that is not clearly specified in advance, or noted carefully as some sort of clearly justified and agreed variation. This applies also to communication practices.

3 Standardisation

In the problematic situation just described, a natural and common reaction is to suppose that the best approach to keeping matters under control is to standardise everything. Many things are standardised already, of course, especially building components and operations. If standard components are being assembled in standardised ways, then it is natural further to suppose that we can standardise any discussion that needs to take place around these. We can have standardised vocabularies, and standardised communication tools. To make sure that there is no uncertainty, we can formalise human roles and relationships. If there is no scope for non-standard actions, then nothing can happen that has not been anticipated, and for which there is not already a defined procedure or protocol. This should mean that quality can be much more effectively controlled, as well as safety, cost, etc. Such standardisation is well under way. International organisations such as ISO have developed standards for data exchange (e.g. STEP, ISO 10303), data modelling (IFC), building construction information (ISO 12006), and many other things. In large building projects, there is the objective of expunging all need for any unexpected human communication on-site; all details should have been determined and prescribed in advance. Any discussion that takes place should itself be subject to standard means of expression and resolution, which can perhaps even be conducted by automated computer agents. Our question in this paper concerns the feasibility of such approaches when confronted by the detail of reality.

As Konrad Wachsmann [12] once put it, "the industrialised process can only have its full effect within a system of all pervasive order and standardisation". But Wachsmann was a pioneer of industrialisation in construction, whose particular systems have not survived. It's certainly at least arguable that his systems failed to survive because they were not sufficiently adaptable. Looking at his USAF Hangar Project, what appears as a modular repeatable system requires nuanced detailing for the repetition to succeed. The problem with imposing a system of all pervasive order is that it assumes things are neat and can easily be kept tidy, and that they exist in a stable environment. But in reality, this is often not the case.

4 Mess

Building sites are messy. One need not spend long looking at them to realise this[1]. Particular topographies and geologies have to be coped with, the weather cannot be controlled, there are delays and other complications that arise from unknown conditions off the site. Although increasingly industrialised, the process is far removed from the operations of sleek robots on an assembly line, partly because the majority of processes that can be industrialised to that extent have already been removed to factory assembly lines and those that remain on site have to cope with the unrelieved particularity of their situation. On the building site, things are not easily kept neat and tidy, so processes need to exist that can be used to move them back towards that state when they drift away (approved drawings), or to legitimise the drift (variations and Architect's Instructions). There are procedures (and much associated documentation) to handle issues like delays, unexpected or inappropriate deliveries, accidents, etc. But all of these require interaction between people, and this alerts us to a somewhat hidden but actually very important aspect of the mess around a building site: *communication* is also messy.

---

[1] Consider, for example, the images and time-lapse videos of a large site available via http://newbuildpics.inf.ed.ac.uk/

The messiness of communication in construction has been discussed elsewhere [8]. There, it is argued that sanctioned and formal modes of communication are inadequate, in practice, for the day-to-day needs of building site reality. The procedures and protocols, documentation and regulation, in fact create a system of a complexity and inertia that would render much of the needed communication impossible. This ultimately undercuts the kinds of standardisation efforts mentioned in section 3 above. The smooth operation of construction projects is dependent on *gaps* and *slippage between* formal communications. This theme is enlarged in McMeel [9], on which we draw to some extent in what follows. It is emphasised that the kinds of issues we have been discussing often demand a *creative* solution. People need the freedom to elaborate their approach in an unanticipated way. Information may need to be created, communicated and manipulated quickly and flexibly. The route to a solution is not provided, or perhaps even accommodated, by the formal mechanisms available, at least without unacceptable difficulty and delay. When these formal communicative systems do accommodate such information it is in danger of being misunderstood and mis-represented as it moves between different interest groups; when they cannot accommodate such information it becomes fragile and in danger of being lost. Within an environment such as the construction site, which is constantly in a fluid state, the perception and validity of specific information that is delivered in quantitative or schematic form from an office environment can accidentally be changed or lost when it enters the site, causing costly delays or confusion.

It may be urged, at this point, that in large and sufficiently well run construction projects these issues do not arise. They are forced out by intensive design and engineering, extensive computerisation prior to site work, and the increasing professionalisation of site management and communication practices. In some, perhaps many, such cases we can concede that unanticipated problems are relatively rare. If so, however, their very rarity may make them more severe, because (by definition) they elude the capacity of the formal mechanisms to resolve them, and informal mechanisms will be minimally available in these cases.

The relevance of a particular piece of information depends not only on who has it and where they have it, but most importantly it can depend on when they have it. Observation of site activities readily yields examples. Information sketched on a plasterboard wall, regarding the position of insulation and the air gap required, can be right beside the location where the detail is applicable, and it is relevant to the location until all the insulation and plasterboard is erected, at which point it will become redundant and will be subsequently plastered over. Similarly, it transpires that often on a building site when vital information is required, staff do not follow the sanctioned channels but rather seek a path of lesser resistance and reach for their mobile phone. The mobile, which site regulations deplore or even outlaw, offers an irresistible immediacy but also an irreplaceable and untraceable informality. Information can be accessed exactly when needed, and it can also be discussed, negotiated and translated without the constraints implied by documentary precision and permanent records.

This affords the potential for swift and simple resolution of disputes. It also allows the participants to take advantage of many of the little features of face-to-face communication that help it run smoothly. These may perhaps be no more than prosodic indications of status and role, but they intersect with rituals and practices on the site that have deep historical roots. Local, creative solutions to problems have always been a feature of construction, and have always depended on people's experience-based understandings of how to work with each other, how to elicit required details, how to promote collaborative focusing on a task. Even though technical design and manufacture has replaced the oral tradition of the master builder and craftsman, the fact remains that each newly developed system brought to a site has to be installed on the basis of hard-won practical knowledge, refined skills, and detailed discussion. From the point of view of the enthusiast for pristine technology and process, these aspects may constitute undesirable "dirt" in the system, but they cannot be abolished. Grease in a machine may appear extraneous and dirty, but its role is nonetheless central in keeping the mechanism working smoothly.

5 Formality and revelation

On the other hand, informal communication runs the risk of unresolved misunderstandings, fails to support retracing of the situation if some problem develops later, and does little to capture successful solutions for future application in similar situations. In a construction project, there is a main contractor, who has the task of coordinating all the other players — this is a difficult job in any circumstances, but it is only made more difficult by the other players communicating informally among themselves without keeping records, something the mobile phone seems to facilitate remarkably well. Overcoming these drawbacks is an objective legitimately pursued by those who introduce formal systems. There are in fact good reasons to frown on the use of mobile phones among construction workers. But what emerges here is clearly a tension between the longer-term needs of the activity overall and the immediate needs of those involved in carrying it out.

In principle, it looks as though we can help with this by creating a system in which everything is very clear and well-defined. We can develop an ontology that covers all of the aspects that might need to be discussed, and make sure that all the terms it uses are well understood by all the participants. This will facilitate clear and unambiguous communication. We can, if necessary, have a hierarchy of ontologies, so that things can be characterised at the appropriate level of detail. Then the use of this system can be incorporated into the processes that are defined for documentation and dispute resolution, etc.

In practice, this continues to fall foul of the problems surveyed in the last section. There is always the potential for people to have differing understandings of concepts included in the ontology, and so great is the diversity of people involved in construction that these differences are bound to emerge. The ontology will be used to define a general interpretation of a term, intended to cover all of its uses; but such generality is inevitably a "lowest common denominator" — that which is in some sense shared by the majority of the known or normal interpretations. We may want people to stick to this, but in practice they cannot, because their own context informs and enriches their natural interpretation of a term, and more importantly this is often a necessary aspect of employing the term effectively in relation to that context. The general interpretation simply lacks the degree of resolution needed to apply it to a specific situation.

One apparent route to a solution is that the ontology can be extended and complicated to an extreme degree, with hierarchical sub-ontologies for all imaginable special cases. Ultimately, this will fail, because it will be unwieldy and there will always arise special cases that have not been imagined. Another approach is recommended here. We can recognise that an ontology is useful as a set of guiding principles to how the domain is structured and where important issues are likely to emerge. We can keep the ontology at a relatively high level of abstraction, but recognise also that differences in interpretation will crop up. In fact, we should embrace these differences and celebrate them as indicators that conceptual clarification is required. Such clarification will proceed by discussion and negotiation, which will open up and relate differing perspectives on the concept. The role of the ontology, on this view, is therefore to identify key terms in the area of activity and to *reveal* differences of interpretation that arise in relation to these, rather than to pretend that there are no differences and force them underground where they are addressed only by the illicit and ambivalent practices that we have mentioned.

6 A related perspective

Work in cognitive science has shown that when people are working together on a communicative task, they tend to develop their own "conventions" about how to use language for that task. This happens very quickly, even in a simple experimental situation where people play a game that involves identifying positions in a small maze drawn on a grid [6]. It happens between pairs, but also spreads to a whole "virtual group" when the members work in successive pairs until each has worked with all the others. (A "virtual group" is defined purely by who has worked with whom, and the group members do not realise that they form a group.) The task is then performed very smoothly by pairs drawn from the group. However, if there is more than one such group they will evolve different conventions, and then if a pair is put together from two different groups the task becomes much more difficult for them. This is a quite robust phenomenon and has been found to occur also with communication using drawings rather than language, where participants evolve and adopt simple but obscure symbols that are very difficult for others to understand [5].

Its relevance to the present discussion is the analogy with groups working together, say on a series of projects for a particular contractor. This sort of group will develop its own communicational styles and shortcuts. Moving from project to project, the group will interact with others, and in each case will be like the people drawn from different groups. Communication will be possible, of course, but it will be less smooth and more prone to error than when communicating within the group.

A key point in this analogy is that adopting the use of an ontology is unlikely to change things radically. It is probably true that adopting a standardised method of referring to locations in the maze would improve inter-group communication in the Garrod et al. maze task, but differences in the uses of this method would still arise and gradually accumulate. Similarly, in the graphical communication situation, standard symbols could be defined, but small differences would appear in the ways these would be drawn, and eventually quite different symbols might well emerge after all.

We could even, in this situation, try to characterise the differences between the groups' communication conventions by describing their divergence from the standardised method. So the standardised method in this situation has a value, but its value once again is to *reveal* the differences between the communicating groups, and to allow us to assess and take account of these.

7 A useful process

Key to the successful revelation and exploitation of these differences is the process of negotiation, which is a very useful process in a number of ways. Communication in construction contexts will often be concerned with the resolution of various kinds of problems, hence processes of problem solving are important. Negotiation is a valuable problem-solving tool.

We have noted that the standardised ontology is necessarily defined at a comparatively general conceptual level. Problems, however, typically arise with issues that are highly specific to some particular instance or exemplar of a general concept. To the extent that approved procedures and protocols are laid down at the general level, it's necessary to bring them to bear on the particular by bringing the latter "under" the generalisation. How one does this is in some sense a matter of perspective: the instance has to be seen as exemplifying the general concept, and there will quite likely be more than one possible way to do that. Here is one source of divergence between individual understandings of a description of the problem situation when general terms are used. A process of negotiation about how the problem is to be described will begin to expose this divergence.

This process might appear to be wasteful, in that it creates multiple expressions of what is supposed to be the same thing. Ambiguity will emerge, there will be unnecessary complexity. However, another way of looking at this is that it creates redundancy, the value of which, within communication, has been explored elsewhere [11]. As in certain kinds of computer systems, this can result in greater "fault-tolerance" in argumentation and a more robust solution in a problem situation. While the process may be somewhat messy, there is a greater chance that an eventual tidying up will not have missed anything important.

But also, in problem solving, there is often a premium on proliferating expressions or representations of the problem. This is obvious in design, for example. Sketches, models, discussions; all are employed as externalisations of thinking that very often give rise to new perspectives, further thinking, and further rounds of representation. This can be extremely fruitful. In collaborative problem solving, of course, these externalisations are critical to ensuring that the collaborators are aligned in their thinking and can maximise the extent to which they aid each other's work. Negotiation is a process that exactly fosters this kind of mutual expression of individual understandings of how the particular can be made sense of in general terms. It is therefore likely to have a strong facilitating effect on problem solving. In the present context, the value of this is to be contrasted with the effect of applying a standard solution and perhaps missing the subtleties of the particular case that may cause that solution to fail.

Again, to return to ontologies, we therefore want to exploit (or perhaps we should say subvert) them in the role of facilitating collaborative problem solving by using them to propose wholly negotiable terms in which to begin the characterisation of a problem situation. At present, this is not something that most developments involving ontologies do much to support.

8 Ontologies and Platonism in the construction industry

There seem to be various things that we might want to use ontologies to help with, e.g.: consistent specification of products; product data exchange; building performance modelling; assembly processes; communication ... — These may not all have the same implications for design and use of ontologies. Current developments tend to focus very much on the first three or four items in this list. There is a strong relationship between ontologies and various efforts to standardise "product models" in the building context, such as the IFCs (Industry Foundation Classes) being developed by the International Alliance for Interoperability (IAI). Whereas data exchange standards for a long time focussed on the syntactic level of how information should be represented, the widespread use of XML now encourages standardisation efforts to shift to the semantic level of the things about which information should be represented. The original philosophical meaning of "ontology", as essentially a description of "what there is" in the real world (to be compared with the common usage of "ontology" as a somewhat arbitrary list of terms to be used in some context), is sometimes thus taken rather seriously, and the ambition may be to create a standard, exhaustive and correct listing of all relevant concepts in the construction domain. Where more limited than this, the restrictions on the ambition seem usually to arise in the extent of coverage of the domain attempted, rather than the nature, status or usage of the representation intended to be achieved.

There are many problems with this general idea, among them that change and innovation are stifled and that varying perspectives cannot be accommodated [10]. For all the reasons we have discussed, these features alone are fatal to any attempt to support communication. This might seem odd, in as much as any ontology ultimately arises from communication. The only access we have to the concepts in play in a domain is overt communication between the players. In the building domain, we study the documents, the drawings, the databases, the responses of experts when questioned — all of these things are communicative artefacts that embody the conceptions we seek to formalise. So one might have thought that any ontology thus derived would surely be well suited for facilitating those same communicative processes. But the error here is an old one. Plato, no less, thought that things in themselves had ideal Forms, and that if we could only represent these Forms we could avoid all the "imperfection", dirt and mess associated with our lowly corporeal being and the inevitable inadequacy of the languages (and other communication systems) that we set up between us. Similarly, both philosophical and technological theorists have sought to abstract from the messiness of language to capture a shared, underlying set of ultimately correct concepts. However, one insight that Plato had, apparently lost on many contemporary ontologists, was that actually representing the Forms would for us always be impossible, since we cannot escape our corporeal being. So any system we set up to try to capture the ultimate nature of things will be at best some sort of rough approximation, subject not only to correction and revision, but also to challenge from alternative conceptualisations offering equally valid representations. This seems a pedestrian observation now, but still it is at odds with many of the objectives of standardisation work with ontologies. And, long after Plato, we have seen the emergence of views, such as those of Wittgenstein [13], implying that communicative practice itself is ultimately "what there is".

Rather than a Platonic vision of construction, then, we urge a view in which "dirt" is in the eye of the beholder (cf. [8]). Douglas [4] illustrates the problematic of dirt as "matter out of place"; if something appears as dirt it is as a result of one's perspective. From another perspective, the same thing might not be considered dirt. Plato thought that everything (on Earth) is dirt, but we allow that some things may be elevated, for certain purposes and in certain contexts, to the more sublime and abstract status of representations that can be treated at least as markers of current agreement to be given priority in discussion. We therefore advocate also a Bakhtin-inspired vision of construction as a carnivalesque activity [8][9] that celebrates dirt, because it is only by contrast with what is otherwise considered dirty that anything emerges as clean, and because at least periodically activities should be promoted in which the clean and dirty are mixed to see whether a more useful perspective on the distinction between them can be found.

9 Dishing the dirt?

Following this line of thought, we would see as valuable a system that collects dirt, where it arises, and instead of disposing of it preserves it, so that it can be dished out in circumstances where it might be useful. If we can retrieve ourselves from metaphor, this amounts to saying that we would like to promote negotiation and discussion of a relatively informal kind, but ideally we will allow this to occur in some environment that facilitates capturing its important content. At some stage, the discussion material might be used to reconfigure the ontological concepts from which it initially derived; at the very least, it will allow some account to be taken of the differing understandings of those concepts that arose in some specific context.

This idea is perhaps not very different from the idea of capturing discussion that surrounds the collaborative development of a conceptual framework, which is something that happens all the time in design. The large body of existing work on capturing argumentation and "design rationale" is therefore highly relevant. Space precludes a detailed summary of this work here. The classic work of Conklin and Begeman [3] on IBIS and gIBIS remains a good starting point; see also [1]. For present purposes, it suffices to emphasise that the point of such a mechanism would be to integrate with the ontology to provide a flexible system that tracks local convention and concept development.

The uses of this kind of material could be various. Sometimes, it would be of very temporary relevance. Like the sketch on the plasterboard wall, it might relate to a quite specific time and place, focusing and refining the ontology for a purpose that simply disappears after a short time and can be forgotten without loss. At other times, it might record a major dispute between, say, a steel erector and a cladding contractor, about the precise description of a framework component. In this kind of case, the outcome could be of lasting relevance for the project in question, and possibly many others. It might then be fed into refinement and development of the ontology itself. It could even contribute to ongoing activities in standardisation working groups such as ISO TC59 (Building Construction) — because standards themselves are, of course, not things that emerge and then are fixed, but are in fact subject to a constant process of change and redevelopment.

10 Ontology mapping

As noted above, ontologies cannot hope to be definitive of the ultimate reality of some domain. And, of course, in practice it very often is the case that several ontologies are developed by different groups or companies, intending to capture broadly the same conceptual field. This is certainly true in construction, where differing perspectives such as those of designer, engineer, services expert, HVAC specialist etc. all import their own concepts. In this situation, one is naturally faced with the problem of how these perhaps rather different ontologies relate. They *should* relate at some level, one supposes, since they are aimed at the same underlying subject matter.

Already, a number of formal approaches have been developed to address this problem, usually along the lines of defining a mapping relation between the ontologies. Systems such as Ontomorph [2] are emerging, which analyse the structure of ontologies to derive morphisms (e.g. homomorphisms) between them at various levels. These systems usually have relatively little to go on, beyond the bare graph-structure of the conceptual network in the ontologies. We do not here envisage the attempt to map an entire ontology onto another, but one might focus on a situation where a particular node is in question, e.g. one is trying to make sense of two intending collaborators who come together to work on a particular aspect of a building, bringing different ontologies with them. In a case like this, it is possible, at least, that the process could usefully be assisted by a record of how this particular node is discussed by the parties, and possibly of how it has been discussed in previous such cases. Whether, and if so how, it might be possible to automate any of this process, we do not here speculate about, but such a system might be valuable purely in support of the negotiation process.

11 Conclusion

We conclude by restating that the use of ontologies in construction should be conceived as the development not of a system that defines the domain in sufficient detail to obviate disagreement and confusion, but rather of a system that attempts to reveal, structure and capture, and thus help to resolve, the disagreement and confusion that is inevitable in such a complex situation. We must remain sensitive to the differing conceptions of different groups or communities of practice acting together on the building site. We should beware of seeking to exclude the "dirt" and sheer messiness inherent in communication, because the pristine is commonly also the sterile (cf. [7]), and because the carnivalesque encounter with dirt is often the stimulus to reconsideration and renovation of tired structures and orientations.

This proposal is obviously not without its difficulties. Capturing rationale and negotiation is notoriously difficult in practice, introduces problematic overheads, and can itself become a focus for unwelcome sanitisation and standardisation. But nothing is gained without cost, and it can hardly be claimed that the straightforward and naïve application of standard ontologies is any less fraught. We hope at least to have provoked the idea that some alternative to that should be sought.

References

1. Buckingham Shum, S. (1996) Design Argumentation as Design Rationale. In *The Encyclopaedia of Computer Science and Technology*, Marcel Dekker Inc: NY, Vol 35 Supp. 20, 95-128.
2. Chalupsky, H. (2000) OntoMorph: A Translation System for Symbolic Knowledge. *Principles of Knowledge Representation and Reasoning: Proceedings of the Seventh International Conference on Knowledge Representation and Reasoning (KR-2000)*, Breckenridge, Colorado, USA, April 2000.
3. Conklin, J. and Begeman, M. (1989) gIBIS: A Tool for All Reasons. *Journal of the American Society for Information Science*, 40, pp 200-213.
4. Douglas, M. (1978) *Purity and danger: an analysis of concepts of pollution and taboo,* Routledge.
5. Fay, N. Garrod, S., Lee, J., and Oberlander, J. (2003) Understanding Interactive Graphical Communication. Paper presented at the Cognitive Science Society conference, Boston, Aug. 2003.
6. Garrod, S. and Doherty, G. (1994). Conversation, co-ordination and convention: An empirical investigation of how groups establish linguistic conventions. *Cognition*, 53, 181-215.
7. Hyde, L. (1998) *Trickster makes this world: mischief, myth, and art.* Farrar, Straus and Giroux, New York.
8. McMeel, D., Coyne, R., and Lee, J. (2005) Talking Dirty: Formal and Informal Communication in Construction Projects. CAADFutures. Vienna.
9. McMeel, D. (2006) Carnival and Construction: Towards a Scaffolding for the Inclusion of ICT in Construction Process. eCAADe, Volos, Greece, Sept. 2006.
10. Ramscar, M., Lee, J, and Pain, H. (1996) A cognitively based approach to computer integration for design systems. *Design Studies* 17, 465-483.
11. Reddy, M. J. (1979) The Conduit Metaphor: a case for frame conflict in our language about language. In Ortony, A. (ed.) *Metaphor and thought,* 284-324. Cambridge University Press.
12. Wachsmann, K. (1961) *The Turning Point of Building*, Reinhold Publishing Corporation.
13. Wittgenstein, L. (1953) *Philosophical Investigations*. Blackwell.

Ontology Based Communications Through Model Driven Tools: Feasibility of the MDA Approach in Urban Engineering Projects

R. Grangel[1], C. Métral[2], A.F. Cutting-Decelle[3]**, J.P. Bourey[4], and R.I.M. Young[5]

[1] Grupo de Investigación en Integración y Re-Ingeniería de Sistemas (IRIS), Dept. de Llenguatges i Sistemes Informàtics, Universitat Jaume I, Castelló, Spain
grangel@uji.es

[2] Institut d'architecture, University of Geneva, Switzerland
Claudine.Metral@archi.unige.ch

[3] Industrial Engineering Research Laboratory, Ecole Centrale Paris, Chatenay Malabry, France
afcd@skynet.be

[4] Industrial Engineering Team, Ecole Centrale Lille, Villeneuve d'Ascq, France
Jean-Pierre.Bourey@ec-lille.fr

[5] Department of Mechanical and Manufacturing Engineering, Loughborough University, Loughborough, United Kingdom
R.I.Young@lboro.ac.uk

Abstract. Enterprises today face many challenges related to lack of interoperability. Enterprise applications and software systems need to be interoperable in order to achieve seamless business across organisational boundaries and thus achieve virtual networked organisations. In this paper, we briefly introduce some key principles of the MDA approach and the role of ontologies in model transformation approaches. Then, we propose a description of the Model Driven Development (MDD) Interoperability Framework. The last part presents a way of applying the MDA techniques to Urban Civil Engineering projects, with the objective of testing the feasibility and relevancy of the approach to this domain.

1 Introduction

Enterprises today face many challenges related to lack of interoperability. Enterprise applications and software systems need to be interoperable in order to achieve seamless business across organisational boundaries and thus achieve virtual networked organisations. IEEE [1] defines interoperability as 'the ability of two or more systems or components to exchange information and to use the information that has been exchanged' [2].

Model-Driven Architecture® (MDA®) [3] is the OMG instantiation of an approach to software development known as Model Driven Engineering (MDE)

** Author to whom correspondence should be addressed.
Email: afcd@skynet.be

R. Grangel et al.: *Ontology Based Communications Through Model Driven Tools: Feasibility of the MDA Approach in Urban Engineering Projects,* Studies in Computational Intelligence (SCI) **61**, 181–196 (2007)
www.springerlink.com

or Model Driven Development (MDD). MDD focuses on Models as the primary artefacts in the development process, with Transformations as the primary operation on models, used to map information from one model to another. There is presently an important paradigm shift in the field of software engineering that may have important consequences on the way information systems are built and maintained [4].

MDD, and in particular MDA, is emerging as the state of practice for developing modern enterprise applications and software systems. The MDD paradigm improves the way of addressing and solving interoperability issues compared to earlier non-modelling approaches. However, developing correct and useful models is not an easy task. We believe that there is a need for an interoperability framework that provides guidance on how MDD should be applied to address interoperability [2].

A key to the success of MDD is the development of ontologies supporting the mapping from one model to another, either at the same level of abstraction or at different levels. Various approaches have been proposed and tested, starting from common ontologies, shared by all the models, to local ontologies, specific to each software. We propose here to discuss the applicability of MDD to the Urban Civil Engineering (UCE) field. This should help to establish the requirements for ontologies to be applied for interoperability of systems commonly used in this domain. In the first section of this paper, we briefly introduce some key principles of the MDA approach and the role of ontologies in model transformation approaches. The following section describes the Model Driven Development (MDD) Interoperability Framework. The last part presents a way of applying the MDA techniques to Urban Civil Engineering projects, with the objective of testing the feasibility and relevancy of the approach to this domain.

2 The MDA Approach

The Model Driven Architecture (MDA) defines an approach to IT system specification that separates the specification of system functionalities from the specification of the implementation of this functionality on a specific technology platform. MDA defines a model architecture through the development of a set of guidelines for structuring specifications expressed as models [5]. The MDA approach and the standards that support it allow the same model functionality to be achieved on multiple platforms through auxiliary mapping standards, or through point mappings to specific platforms. It also allows different applications to be integrated by explicit relations between their models, thus enabling the integration, the interoperability and the evolution of supporting systems.

2.1 Basic Concepts

MDA begins with the idea of separating the specification of the operation of a system from the details of the way the system uses the capabilities of its platform [3]. MDA provides an approach for, and enables tools to be provided for:

specifying a system independently of the platform that supports it, specifying platforms, choosing a particular platform for the system and transforming the system specification into one for a particular platform. The primary goals of MDA are portability, interoperability and reusability. The MDA concepts are presented in terms of existing or planned system. This system may include anything: program, single computer system, some combination of parts of different systems, federation of systems each under separate control, people, enterprise, federation of enterprises. The discussion focuses on the software tools within the system. A model of a system is a description or specification of that system and its environment for a given purpose. A model is often presented as a combination of drawings and text. The text may be expressed in a modeling language or using a natural language. MDA is an approach to system development. It increases the power of models. It is model-driven because it provides a means for using models to improve the understanding, design, construction, deployment, operation, maintenance and modification.

The architecture of a system is a specification of the parts and connectors of the system and the rules for the interactions of the parts using the connectors [6]. MDA prescribes certain kinds of models to be used, defining a hierarchy of models from three different points of view: the Computation Independent Model (CIM), the Platform Independent Model (PIM), and the Platform Specific Model (PSM) [5].

The computation independent viewpoint focuses on the environment and the requirements of the system; the details of the structure are hidden or not yet defined. The platform independent viewpoint focuses on the operation of a system while hiding the details necessary to a particular platform. A platform independent view shows the part of the complete specification that does not change from one platform to another. A platform independent view may use a general purpose modelling language, or a language specific to the area in which the system will be used. The platform specific viewpoint combines the platform independent viewpoint with an additional focus on the detail of the use of a specific platform by a system.

2.2 Model Transformation

Model transformation defines the process by which a model is converted to another model of the same system. The MDA process shows the role the various models (CIM, PIM, and PSM) play within the MDA framework. A transformation tool takes a CIM and transforms it into a PIM. A second (or the same) transformation tool transforms the PIM into PSM. The transformation tool takes one model as input and produces a second model as output. Generally speaking, a transformation definition consists in a collection of transformation rules, which are unambiguous specifications of the way (a part of) one model can be used to create (a part of) another model. Based on those observations, it is possible to define transformations, transformation rules and transformation definitions [7], such as:

- **transformation:** automatic generation of a target model from a source model, according to given definitions;
- **transformation definition:** set of transformation rules that together describe how a model can be transformed from the source language into a model in the target language;
- **transformation rule:** description of how one or more constructs in the source language can be transformed into one or more constructs in the target language.

The main transformations are [5]:

- **PIM to PIM:** transformation used when models are enhanced, filtered or specialised during the development life cycle without needing any platform dependent information. One obvious mapping is the analysis of design models transformations. PIM to PIM mappings are generally related to model refinement.
- **PIM to PSM:** used when the PIM is sufficiently refined to be projected onto the execution infrastructure. The projection is based on the platform characteristics. Describing these characteristics is done using a UML description (and eventually a profile for describing common platform concepts). The translation from a logical component model to a commercial existing component model is a kind of PIM to PSM mapping.
- **PSM to PSM:** transformation needed for component achievement and deployment. For example, component packaging is done by selecting services and preparing their configuration. Once packaged, the components delivery could then be done by specifying initialisation data, target machines, container generation and configuration, etc. PSM to PSM mapping are generally related to platform dependent model refinement.
- **PSM to PIM:** transformation required for abstracting models from existing implementations in a particular technology into a platform-independent model. This procedure often resembles a 'mining' process generally hard to make fully automated. It may be supported by tools. Ideally, the result of this mapping will match the corresponding PIM to PSM mapping.

There are several kinds of model transformation approaches: an interesting classification, based on design features, has been developed by Czarnecki and Helsen [8].

2.3 Use of Ontologies in Model Transformation Approaches

A key-feature of the MDA approach is 'interoperability'. There exist various definitions of interoperability. According to the Oxford Dictionary, interoperable means 'able to operate in conjunction'. The word 'interoperate' also implies that one system performs an operation on behalf of another system. From a software engineering point of view, interoperability means that two co-operating software systems can easily work together without a particular interfacing effort. It also means establishing communication and sharing information and services between software applications regardless of hardware platform(s). The interoperability is considered achieved if the interaction can, at least, take place at the three levels: data, application and business process with the semantics defined in a business context. The MDA approach analyses interoperability issues from the point of view of system architecture and platforms, through transformation approaches between the CIM, PIM, and PSM models [4].

Ciocoiu et al. [9] discuss two approaches for the use of ontology in interoperability. First, the standardisation approach, wherein all are encouraged to use a common, shared, standardised ontology for their enterprise applications. Such an approach has, however, been deemed impractical. Secondly, the interlingua approach, wherein ontology is used as an interlingua and interpreters/translators are written from the ontology to/from the software applications.

The single ontology approach works well if an ontology is to be designed and modelled from scratch. However, the ontology is usually limited to the purpose of its application. That is, it has limited reusability outside the scope of its application. Multiple local ontologies are generally applicable and reusable but difficult to implement practically. Thus, as Ciocoiu et al. [9] propose, the Interlingua approach is a middle road approach. It tries to overcome the applicability problems of a single ontology while keeping translation problems at a manageable level. They propose to have a shared ontology and to use it as an Interlingua for translating between communicating systems (much like P2P interoperability). It allows the communicating systems to have their own local ontologies. This approach also has the advantage that the networked system can be easily extended to include other systems as well, without having to create a large number of mapping relations. In practice, the implementation of point to point translators between every pair of applications requires N(N-1) translators for N applications. Further, if any new application is to be added, N new point-to-point translators are required. With the use of a common shared ontology (which provides all the applications with a common set of terminology and semantics), the number of point to point translators required is reduced to N translators, from each application to the common translator. The drawback with this approach is that all the interoperating applications need to understand the formalism/ontology representation language of the shared ontology. Or else they need to be able to interoperate or translate within the different ontology representation formalisms like KIF, DAML, RDFS to name a few. Also, a minimum consensus view for the common ontology needs to be reached through mutual agreements or standardisation processes.

Ontology integration is an important task to achieve interoperation between systems using different ontologies. Sowa [10] defines ontology integration as the process of finding commonalities between two different ontologies O and O' and deriving a new ontology O". The new ontology O" may replace O or O', or it may be used only as an intermediary between a system based on O and a system based on O'. Three general approaches for combining distributed heterogeneous ontologies can be distinguished [11]:

- **Ontology Inclusion** in which the source ontology is simply included within the target ontology. All definitions from the source ontology are included within the target ontology and it is not possible to include only parts of the source ontology.
- **Ontology Mapping or Alignment** which is the weakest form of integration. An alignment is a mapping of concepts and relations between the two ontologies that preserves the partial ordering by subtypes in both ontologies. A concept in one ontology can be mapped to a concept in the other ontology with an equivalence or an inclusion link. The mapping may also be partial: there could be many concepts that have no equivalents in the other ontology. The drawback of this approach is the need for flexible mechanisms for transformations.
- **Ontology Merging** using Mediators which is the most complex approach combining several data sources into a single integrated ontology through the use of a mediator to answer queries [12].

Purely manual approaches are insufficient to support large or dynamic systems interoperability. Therefore partially or fully automated applications have been developed for the merging, mapping or alignment of ontologies. Many algorithms (e.g., PROMPT [13], GLUE [14], Ontrapro [15], OLA [16], FOAM [17]) have been proposed. Semantic similarity measures play a central role in ontology matching with some differences between methods. For example, while many mapping systems incorporate only a single similarity function to determine if two concepts are semantically related, GLUE utilises multiple similarity functions to measure the closeness of two concepts depending on the purpose of the mapping.

MAFRA (Mapping FRAmework [18]) is another ontology mapping methodology. MAFRA creates a Semantic Bridging Ontology (SBO) that contains all concept mappings and associated transformation rule information. Given two ontologies (source and target), this approach requires domain experts to examine and analyse the class definitions, properties, relations and attributes to determine the corresponding mapping and transformation method. Therefore, SBO serves as an upper ontology to govern the mapping between the two ontologies.

Evaluation of the performance of ontology matching algorithms is often made on the basis of the usual precision and recall measures. However, these measures do not discriminate accurately between methods which do not provide exact results. However, when the alignment results have to be screened by humans, this is an important need. A generalisation of the usual approach is proposed in [19]. This approach keeps the advantages of usual precision and recall but helps discriminating between alignments by identifying near misses from completely wrong correspondences.

3 Application: Model Driven Development (MDD) Interoperability Framework

Why is interoperability so a big issue in software architectures and frameworks? From a system oriented perspective a software system is a conglomerate of subsystems communicating internally and with environment via well defined interfaces. Additionally software systems are designed to provide specific business functionalities. Interfaces in terms of functionality, protocols and signatures are described by models, languages and additional text documents. The challenge of interoperability is induced by the mismatch of these interfaces in multiple senses. Computer can match these interfaces and ensure consistency formally but lacks mapping the interface semantics. The challenge of interoperability is to address the semantics of system specifications in context of business applications [4].

The reference framework described in this section covers a simplified architecture model and its relationship to systems and enterprise models integrated in the context of the enterprises. Modelling is one of the most important tools for software engineering and system design - particularly for Model Driven Development: modelling and model management are key-drivers for interoperability.

The starting point for the INTEROP Interoperability Framework [4] originated in the IDEAS Project [20]. IDEAS analysed interoperability aspects from an enterprise view (i.e. between two or more enterprises), from an architecture & platform view (i.e. between two or more applications/systems), and from ontological views (i.e. interoperability semantics) The Enterprise view is separated into Business issues and Knowledge issues. Based on the ATHENA Framework [21], the INTEROP Project has lead to the development of three views of an enterprise: conceptual, technical and applicative view. Those views are used to provide reference models for integration. We focus in this paper on the conceptual view, for further information see [4].

The conceptual view of an enterprise has been developed from a MDD point of view focusing on the enterprise application software system. A Computation Independent Model (CIM) corresponds to the view defined by the context viewpoint. It captures the business context of the software system. A Platform Independent Model (PIM) and a Platform Specific Model (PSM) are both computational dependent with respect to the software system, the difference being that the PIM is independent of an execution platform while the PSM is not. The models at the various levels may be semantically annotated using ontologies to achieve a mutual understanding whatever the levels. The use of ontologies helps in doing model transformations and mappings between models.

The conceptual integration focuses on concepts, meta-models, languages, and model relations. It provides a foundation for systemising various aspects of ICT model interoperability. Interoperability issues occur both within the company (vertical integration) and between companies (horizontal integration) (see Fig. 1). The primary focus is on the horizontal and vertical integration issues that take place between enterprises (interactions and collaborations). The interoperability patterns applied between companies (inter-) can be recursively applied to solve interoperability issues between business units within a company (intra-).

Model mappings can be defined using meta-models and ontologies. The focus of the reference model is on the horizontal integration. Vertical integration is addressed by MDD and ADM (Architecture Driven Modernisation). Emphasis is on model mapping, synthesis and development with respect to model integration. The use of a reference ontology for semantic annotation of models helps to achieve this integration. Generic or domain-specific interoperability patterns can also be used. Models are used to describe different concerns of the software system.

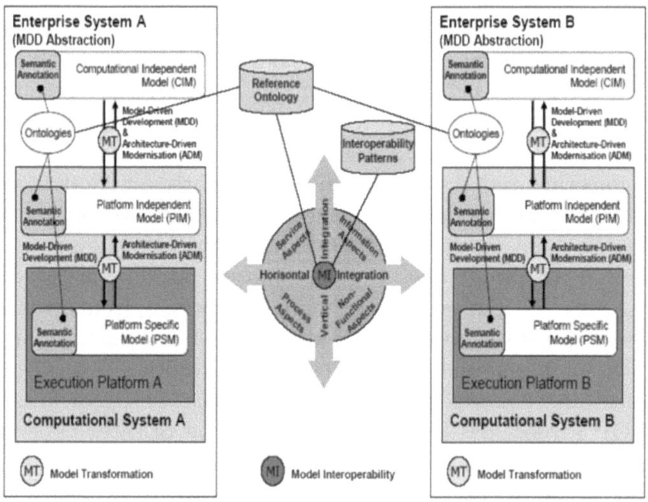

Fig. 1. Reference model for conceptual integration [4]

Model-driven approaches to generating software applications provide many advantages by improving portability, interoperability and reusability through the architectural separation of concerns. It is interesting to see to what extent those concepts can be applied to the ontologies met in Urban and Civil Engineering projects. A first approach is proposed in the following section.

4 Application of MDA Techniques to the Management of Ontologies in Urban Engineering Projects

In this section, we present two examples of ontologies developed in the domain of urban engineering. The need for ontology-based model transformations comes from the differences in the ontologies on which the different applications are built. A first description of this kind of transformation is then provided. This section ends with the presentation of some benefits that can be expected from this approach.

4.1 Ontologies in Urban Engineering: Examples - Role of the Ontologies

Mobility and travel ontology. The objective of this ontology is the integration of concepts coming from a street planning ontology [22]. This work has been split into three phases: construction of a sample of concepts, search of the set of associated definitions and implementation of the semantic network.

To obtain a sample of concepts of the field, various alternatives were possible. Our choice was carried out on many concepts through a questionnaire sent to experts of the domain, in order to identify the concepts interesting the users (given by the experts), in order to facilitate afterwards the future tests of the base, but also to increase the interest carried out with this tool. This method enabled us to collect around fifty concepts in the field of mobility and travel.

The second phase consisted in gathering a set of definitions concerning all the new concepts listed. Some of them had got, since the beginning, several definitions whereas others, on the contrary, missed precise details. It was thus necessary to carry out a work of selection and complementation. Then, each of these definitions was organised according to the existing structuring of the base, that is sorted out according to its specialisation and its activity field.

Based on a sample of concepts and definitions, the organisation in semantic networks began, supporting the relations defined in the existing base. It was necessary to simplify the use of the existing relationships in the semantic network, both by classifying them to remove ambiguities, but also by generalising them, to facilitate their re-use and to avoid their multiplication. In order to integrate this field into the existing model, we selected a certain number of footbridging concepts, in order to make it possible to connect the new semantic network to the existing network.

Figure 2 shows an excerpt of the ontology under the form of a semantic graph. It contains at the same time concepts from street planning and mobility. A relation represents a direction of reading: the relation 'is located on' between the concepts 'Horizontal sign' and 'Pavement' is read as: Horizontal sign 'is located on' the Pavement.

Regulations and Guidelines Ontology. Another example is taken from [23], where the author develops ontological constructs for representing regulations and guidelines with geography. We take here the example of a regulation about prohibited actions. Actor have capabilities to perform certain actions. Choices of actions are however restricted both by norms as well as regulation. However actor intends to perform an action either in relation to assets or activities. The following example illustrates the idea of a regulator prohibiting a type of regulated from taking a certain action, an investment within the city limits. Figure 3 explicitly includes jurisdiction as a part of the antecedent statements to illustrate the idea. Note the explicit reference of location attribute of the investment

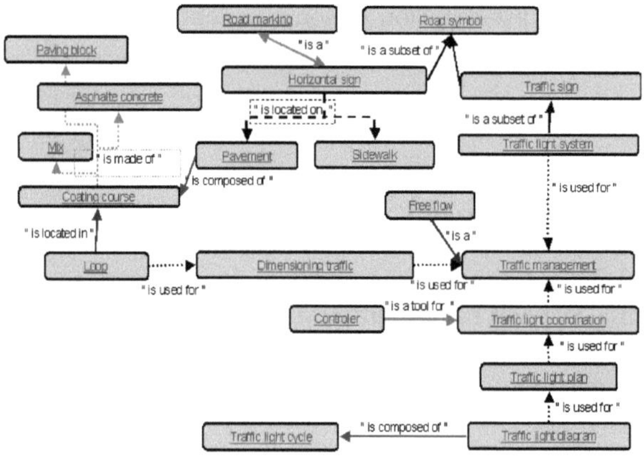

Fig. 2. Excerpt of the ontology [22]

to the spatial limit attribute of the jurisdiction It shall be unlawful for any person within the city to erect an electrified wire fence of any sort (adapted from Urbana Municipal Code 1998 [24]).

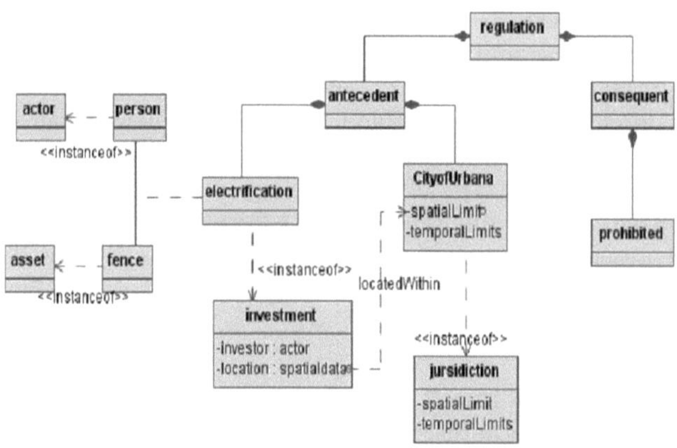

Fig. 3. Example of regulation about prohibited actions [23]

More generally, a generic urban development application can make use of several ontology-based software tools. Concepts about (footprints of) buildings in the GIS ontology could be extended by the urban development application to hold more information than available in the initial GIS system. By extending the GIS ontology, use of GIS data such as the geometry of the footprints of buildings is assured. A water usage prediction web service/application could extend the building concept as well with properties such as waterusage, amount of showers, building type, garden size, etc. Using this information the water service can estimate a water usage per dwelling. Another application capable of calculating energy consumption can extend the ontology with energy slots and relationships such as windowtypes, walltypes, etc. Figure 4 shows a network of such software applications [25]. When the ontologies are interoperable, collaboration is made possible among the partners of the urban engineering project.

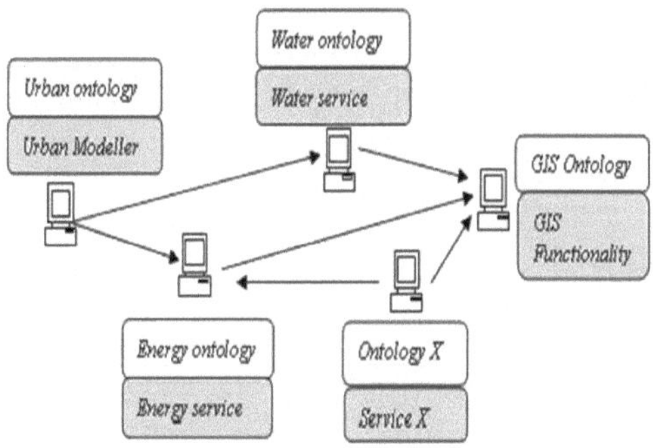

Fig. 4. A conceptual architecture based on a network of software applications used for urban development [25]

All those ontologies need to be aligned to preserve the interoperability of the whole system: in the MDA approach, this alignment is made through ontology mapping: concepts can be related concepts from both ontologies using additional knowledge, i.e. mapping rules. These rules can be used to find concepts in the other ontology that correspond to concepts used. An example is a unit conversion service which can translate imperial units to metric units [25]. More generally, model transformations are necessary to perform more complex translation mechanisms.

4.2 Model Transformations

A real test case has been developed by the Task Group 2 (TG2) of the INTEROP Project, with the aim of providing a model-driven method that could guide enterprises in using enterprise models to generate Enterprise Software Applications (ESA) as automatically as possible, and more particularly to parameterise Enterprise Resource Planning (ERP) systems [26]. This research focused on the difficulties enterprises face to generate ESA from enterprise models, and how a model-driven approach could be a useful way to solve them, while enabling at the same time to achieve interoperability. The TG2 also developed different enterprise models [27] for a case study, at different levels of abstraction and a first set of experiences on model transformation at the vertical level. Based on this work, a first method to parametrise ERP from enterprise models using model transformation in a vertical approach has been proposed [28]. Figure 5 shows the evolution of the initial idea of TG2 towards the use of only one ontology model. At this stage the objective of the work is to provide an interoperability model (meta-level) that can be connected to the parameterisation of ERPs using MDI.

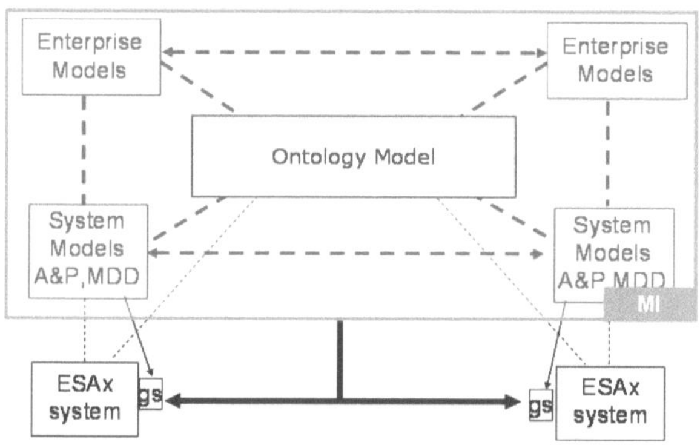

Fig. 5. Model Driven Interoperability: vision targeted by the TG2 [28]

4.3 Benefits of the Approach for UCE Projects

An important feature of urban engineering projects is the importance of the coordination activities. Schmidt [29] suggests the notion of coordination mechanism based on protocols and artifacts. In this domain, the urban project, in the early construction stage, is precise concerning the objectives to be met with. Objects to be built are described in artifacts (plans, technical documents, etc.). But it is less precise concerning the methods and procedures to be used (i.e. time control or task organisation). Planning activity has to give a temporal space and a logical sequence to 'building task' execution. However, this artifact is under-defined. Moreover, it hardly takes into account changes happening during the project. Informal relations between actors are the unique solution to adjusting tasks to the major stages planned [30].

Complexity of the projects and associated cooperative practices lead to particular coordination modes. This production system appears today as well balanced. But there are some dysfunctions which reduce global quality of cooperative activities and then of the architectural object itself: information overload, unlinked information, difficulty in tracing events, risk of redundancy and contradiction between documents, lack of coherence or sometimes absence of information.

New methods and new tools have been developed for some years in order to take into account these limits of coordination. They have been developed to assist the design stage, construction stage or both. Representing the complexity and the particularities of the domain is the first step towards propositions for new assistance tools for cooperation. Model Driven Engineering can be considered as one of those tools.

5 Issues - Perspectives: towards a Meta-model Approach in Urban Engineering

The meta-modelling approach was described by [31] and used in the standard MOF (Meta Object Facility, see Fig. 6). It is proposed by the OMG [32].

For Kubicki et al. [30], and applied to the construction sector, the definition of a meta-model allows to highlight essential abstract concepts to describe context of cooperation in different domains. These 'meta-concepts' of the meta-model (M2 level) are instantiated in specific cooperation models (M1 level): building construction activity context model, meeting-report model, project management model or in other domains such as software engineering. An approach suggested by the authors consists in defining a relational cooperation meta-model that takes into account the existing relations between the elements of a project.

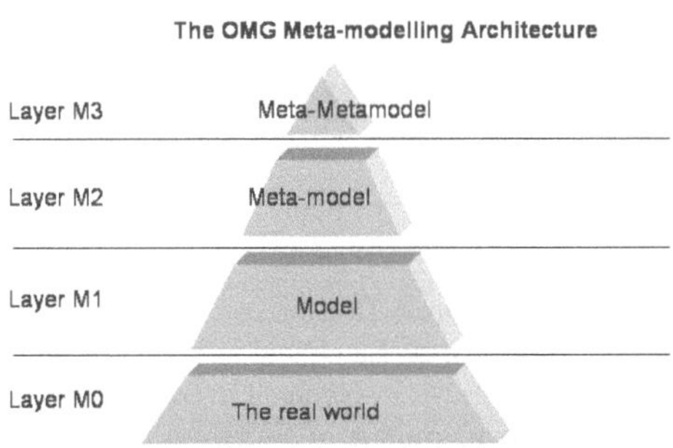

Fig. 6. The OMG Meta-modelling Architecture [32]

To model the activity in a building construction project the authors suggest an approach from the point of view of cooperative activities between actors (i.e. exchanges or dependencies). The context of cooperative design and construction activities represents relations and interactions between the actors, their activities, the documents they produce and the object of the cooperation:

- **Activity (M2):** the activities inside a project have several 'scale' levels: project, phase, and task. They should be explicit (building task) or implicit (request between two actors).
- **Actor (M2):** in a project, each actor has a limited capacity of action and restricted decision-making autonomy. The actor acts inside the activities that constitute the project, gives an opinion, and keeps up a relationship with the environment while collaborating with other actors and producing documents. An actor often works in a group.
- **Document (M2):** a document represents a professional 'deliverable' part of a contract. A document is an aggregation of files manipulated through an operating system. A document can group several other documents. Finally, actors generate documents during activities.
- **Object (M2):** the realisation of the object is the goal of the cooperation project. An object could comprise other objects (group of objects).
- **Relationship (M2):** a relationship identifies a type of link existing between two elements. There are several relationships.

We are currently making investigations to analyse to what extent those concepts proposed for construction projects are applicable to urban engineering projects. This approach looks promising and would enable a more generic way of considering these projects. Also, it would facilitate a common processing of multi-cultural urban projects.

Acknowledgments

The INTEROP project mentioned in this paper is funded by the EC within the 6th FP, INTEROP NoE, IST-2003-508011 [33]. This work was also funded by CICYT DPI2006-14708, CICYT DPI2003-02515.

References

1. IEEE: IEEE Standard Computer Dictionary: A Compilation of IEEE Standard Computer Glossaries. Institute of Electrical and Electronics Engineers. (1990)
2. Elvesæter, B., Hahn, A., Berre, A.J., Neple, T.: Towards an Interoperability Framework for Model-Driven Development of Software Systems. In: 1st International Conference on Interoperability of Enterprise Software and Applications (I-ESA'05). (2005)
3. OMG: MDA Guide Version 1.0.1. Object Management Group. Document Number: omg/2003-06-01 edn. (2003)

4. Berre, A., Hahn, A., Akehurst, D., Bezivin, J., Tsalgatidou, A., Vermaut, F., Kutvonen, L., Linington, P.: Deliverable D9.1: State-of-the art for Interoperability architecture approaches, Model driven and dynamic, federated enterprise interoperability architectures and interoperability for non-functional aspects. Technical report, INTEROP-D9 (2004)
5. OMG: Model Driven Architecture (MDA), Architecture Board ORMSC. Object Management Group. Document Number: ormsc/2001-07-01 edn. (2001)
6. Shaw, M., Garlan, D.: Software Architecture: Perspectives on an Emerging Discipline. Prentice Hall (1996)
7. Kleppe, A.G., Warmer, J., Bast, W.: MDA Explained: The Model Driven Architecture: Practice and Promise. Addison-Wesley Longman Publishing Co., Inc., Boston, MA, USA (2003)
8. Czarnecki, K., Helsen, S.: Classification of Model Transformation Approaches. In: OOPSLA'03 Workshop on Generative Techniques in the Context of Model-Driven Architecture. (2003)
9. Ciocoiu, M., Gruninger, M., Nau, D.: Ontologies for Integrating Engineering Applications. Journal of Computing and Information Science in Engineering **1** (2000) 12–22
10. Sowa, J.: Glossary. http://www.jfsowa.com/ontology/gloss.htm (2006) Accessed November 2006.
11. Maedche, A., Motik, B., Stojanovic, L., Studer, R., Volz, R.: Ontologies for Enterprise Knowledge Management. Journal IEEE Intelligent Systems **18(2)** (2003) 26–33
12. CNR-IASI: Deliverable D8.1: State of the art and state of the practice including initial possible research orientations. Technical report, INTEROP-D8 (2004)
13. Noy, N., Musen, M.: SMART: Automated Support for Ontology Merging and Alignment. In: Twelfth Workshop on Knowledge Acquisition, Modeling, and Management. (1999)
14. Doan, A.H., Madhavan, J., Dhamankar, R., Domingos, P., Halevy, A.: Learning to Map Ontologies on the Semantic Web. VLDB journal **12 (4)** (2003) 303–319
15. Ashpole, B.: Ontology translation protocol (ontrapro). In Messina, E., Meystel, A., eds.: Proceedings of Performance Metrics for Intelligent Systems (PerMIS'04). (2004)
16. Euzenat, J., Valtchev, P.: Similarity-based Ontology Alignment in OWL-lite. In: Proceedings of the 15th ECAI. (2004) 333–337
17. Ehrig, M., Staab, S.: QOM - Quick Ontology Mapping. In: Proceedings of 3rd ISWC. (2004)
18. Maedche, A., Motik, B., Silva, N., Volz, R.: MAFRA - A MApping FRAmework for Distributed Ontologies. In: EKAW'02: Proceedings of the 13th International Conference on Knowledge Engineering and Knowledge Management. Ontologies and the Semantic Web. (2002) 235–250
19. Ehrig, M., Euzenat, J.: Relaxed Precision and Recall for Ontology Matching. In: K-CAP'05, Workshop Integrating Ontologies. (2005) 25–32
20. IDEAS: Deliverable D1.1 part C: SotA on Architecture & Platforms. Technical report, IDEAS (2003)
21. ATHENA: Advanced Technologies for interoperability of Heterogeneous Enterprise Networks and their Applications IP (IST-2001- 507849). http://www.athena-ip.org (2006)
22. Teller, J., Keita, A., Roussey, C., Laurini, R.: Urban ontologies for an improved communication in urban civil engineering projects. In: SAGEO. (2005)

23. Kaza, N.: Towards a data model for urban planning: ontological constructs for representing regulations an guidelines with geography. Technical report, Departement of Urban & Regional Planning, University of Illinois (2004) Project report
24. Urbana: The Urbana Zoning Ordinance, City of Urbana, IL (2003)
25. Schevers, H., Trinidad, G., Drogemuller, R.: Towards integrated assessments for urban development. ITcon Journal **11** (2006) 225–236
26. Bourey, J., Grangel, R., Doumeingts, G., Berre, A.: Deliverable DTG2.2: Report on Model Interoperability. Technical report, INTEROP-TG2 (2006)
27. Doumeingts, G., Berre, A., Bourey, J.P., Grangel, R.: Deliverable DTG2.1: Report on Model establishment. Technical report, INTEROP-TG2 (2006)
28. Grangel, R., Bourey, J., Berre, A.: Solving Problems in the Parametrisation of ERPs using a Model-Driven Approach. In: 2nd International Conference on Interoperability for Enterprise Software and Applications (I-ESA'06). (2006)
29. Schmidt, K., Simone, C.: Coordination Mechanisms: Towards a Conceptual Foundation of CSCW Systems Design. Computer Supported Cooperative Work: The Journal of Collaborative Computing **5** (1996) 155–200
30. Kubicki, S., Bignon, J., Halin, G., Humbert, P.: Assistance to building construction coordination - towards a multi-view cooperative platform. ITcon Journal **11** (2006) 565–586
31. Sprinkle, J., Ledeczi, A., Karsai, G., Nordstrom, G.: The new metamodeling generation. In: IEEE Conference and Workshop on Engineering of Computer-Based Systems. (1999)
32. OMG: A proposal for an MDA Foundation Model. Object Management Group. Document Number: ormsc/2005-04-01 edn. (2005)
33. INTEROP: Interoperability Research for Networked Enterprises Applications and Software NoE (IST-2003-508011). http://www.interop-noe.org (2006)

Modelization of the Conception and Conception of the Model in Architecture

Emmanuelle Pellegrino

CRAAL
Centre de Recherche en Architecture
et Architecturologie
Chemin de Rovéréaz 11
CH-1012 LAUSANNE
empellegri@bluewin.ch

Abstract. In the objective to model the design of the city we develop an ontology which supports a CAD software intended for architects; it understands the architectural concepts like rules making possible a reasoning in a process of project, at the interface between two complementary operations, classification and composition. We propose then to go beyond the limits of an ontology conceived as a tool based on the logic of classes to build an intelligent ontology conceived as a tool based on logic of proposals.

Keywords: classification, composition, interpretation, pɒcess, project, semiotic, model, rules.

E. Pellegrino: *Modelization of the Conception and Conception of the Model in Architecture,* Studies in Computational Intelligence (SCI) **61**, 197–210 (2007)
www.springerlink.com

1. Introduction

These works concern the modelization of the architectural conception processes. We aim for two objectives. The widest one is the development of an intelligent system for architecture project teaching. This system is based on knowledge produced, using semiotics, with the aim to develop a software of computer-aided design intended for architects. The understanding of the way of creating the project and the modelization of the various stages of its process of design become a support for the teaching of the theory of the project. Such a software doesn't intend to replace the designer; it puts at his disposal a vast network whose nodes question him on the future development of his project and guide him to materialize his choices. A designer can find there his own conception of the model, and build his own language.

The most restricted objective is to build an ontology which allows actors belonging to various professions to communicate on the basis of a common vocabulary. According to the point of view of an actor, the same concept can cover with completely different realities.

2. Short history

This research emerges from the continuity of a line of works started in 1994, initially under mandate of the FNSRS [01], in collaboration with mathematicians and data processing specialists. It was a question of formulating the bases to pass from a drawing software conceived as a material design tool to a CAD software conceived as a conception design tool. A double problem arose. First, which information were relevant for such a software, to be able to pass from a request formulated in natural language, in terms of rooms and elementary relations, to a complex composition, formulated in geometrical language. Second, which form of these data is most appropriate for a computer implementation.

A semiotic approach allows us simultaneously to solve both aspects of the problem. Architecture can't exist without forms, its object is the form. Semiotics is defined as the theory of the forms that produce meaning; applied to architecture, it studies the processes by which the meaning of the form is produced. The characteristic of any language is to produce meaning. We took as starting point the definition of the sign of Saussure; for him the meaning emerges from the arbitrary relation of a signified with a signifier [02]. The signified is connected with the form of the content of the building, form that envelops the content of the rooms of the building and the relation between contents; it thus refers to a problem of classification. The signifier is linked with the form of the container of the building, form which gives a geometry, in the sense that these rooms and their relations will take a global form; it thus refers to a problem of composition. Seeing that any translation supposes the transformation of equivalent data, coding architecture as a language allows then a translation in a data-processing language.

In our research, architecture is not a product determined by a geographical and cultural context as it would be for a geographer for whom qualities of various materials result from local conditions. In contrast, to equivalent climatic data can correspond very different architectures as well as dissimilar climatic data can generate similar architectures. Producing meaning, architecture is the fruit of a free will; for the same problem, there is thus a multiplicity of possible answers. With determination, it substitutes interpretation, with passivity invention.

Architecture uses a geometrical language to communicate. The words of the architect are squares, rectangles, figures he assembles in a composition full of meaning. Each architect speaks his own language; a language that can be described in a formula.

Language = {[diagram x metric) x euclidean geometry] x style}

Discourse = context x {[(diagram x metric) x euclidean geometry] x style}

Seeking to describe a structure as well as a process, we initially highlighted the structure of the architectural language in various authors. Through the study of villas of various Masters, we modelized styles, manners. From a small number of buildings considered as texts, texts with unknown vocabulary, style and grammatical rules, we sought to codify the architectural language, so as to be able 'to speak' architecture by producing new texts. Our approach to the problem, from a linguistic analysis, has consisted to highlight a common structure to every analyzed language. This structure defines three processes of design, the first one iterative, the second one operational and the third one normative. The iterative process isolates and reproduces the form of a stable geometrical entity. The operational process is made of geometrical operations allowing to assemble stable components; these operations are translation from rhetorical figures of the natural language into architectural language. The normative process develops a range of rules to prepare the entry of the context and to produce different variants according to the characteristics of a place.

The content designed in each one of these three processes varies according to Masters; articulated in the relation between these three processes, they form the semiotic model from which it's possible to produce variants in a certain context.

2.1. GED (Graph Editor)[1]

Seeing that the work of an architect consists in passing from classification to composition, to transform rooms and relations into a geometry, we needed a tool to help us to know and measure data relating to classification as well as composition. We developed a software able to carry out calculations on graphs. In the principle, a plan of architecture can be translated into diagrams modelizing various of its dimensions, as for example connexity relations, adjacency data or order between rooms, but also other dimensions, as for example the path of the fluids or the degree

[1] Software made by the CRAAL, with the cooperation of D. Coray, J. Colinge, E. Nanchen and K. Ben Saci.

of privacy of the rooms. These diagrams can be in their turn translated into graphs. Values, modelizing dimensions such as degrees of opening, can be registered on the arcs or on the vertice of a graph.

GED measures the distance separating each vertex from every other, then it translates the results into numerical values. It can take into account the classification of the vertice, that's to say the relative position of the ones compared with the others; it can also take into account 'colors', that represent a particular encoding of properties assigned to vertice or arcs of a graph. GED allows then to compare various graphs. One can thus compare the level of depth of various graphs representing a modelization of various aspects of a plan.

This software allows moreover to calculate properties of a graph. One can calculate the valence, i.e. the vertex or the vertice which articulate(s) the greatest number of relations. The rate of eccentricity, i.e. the diameter and the ray of a graph, by measuring the distance separating each vertex from every other. This rate informs about the most central and peripheral vertice. The moment of inertia of a graph, i.e. its point of equilibrium, by taking into account, or not, the surfaces of the vertice. Lastly, one can more calculate the group of automorphism, by counting the number of permutations between the vertice of a graph that leave the graph unchanged. According to the results, the designer will be able to choose such or such other solution by comparing them using numerical indicators.

2.2. TOP (Taxis Oriented Project)[2]

We developed a second software, TOP (Taxis Oriented Project), that treats of questions relating to the geometrical composition. Starting from a stable element, using geometrical operations, TOP composes the figures of an architectural composition.

Architecture concerns thus at the same time a problem of classification, solved by GED, and a problem of composition, solved by TOP. The way of classifying vertice has effects on the setting in geometry and vice versa. At the moment both software are connected in one-way; it's possible to pass from a composition to a classification, (from TOP to GED). We are actually developing the opposite operation, that's to say the way from conceptual graphs to geometrical compositions. As a building is an answer to a problem, a problem formulated through a request and in a certain context, it's not any more a question of modelizing a problem already solved by the interpretation of 'this request' in 'this context', but to bring a solution to a new problem, by successive approaches, exactly as would do an architect. Thus now arises the difficulty of the invention of a project.

---

[2] Software made by the CRAAL, with the cooperation of D. Coray, S. Cirilli, E. Nanchen and K. Ben Saci.

3. Objectives: from an elementary to an erudite ontology

For this reason we intend on the one hand to develop TOP and GED, so that they can interact. But if GED is able, starting from a plan carried out with TOP, to build a graph and to analyze some of its properties, this being, it doesn't tell us the rules according to which we could classify the vertice; GED is an instrument that doesn't suggest an interpretation of its analysis. And if TOP offers a support to draw a basic geometrical entity, and to complexify it, it doesn't tell us which operations to realize for which result. This is why, in addition, within the framework of a project financed by the COST [03], we seek to build an ontology, located at the interface between TOP and GED; an interface which allows to articulate these complementary operations, classification and composition, in any process of architectural project. This ontology informs us on ways of classifying and ordering elements; it also informs on ways of dimensioning and giving forms to classified elements.

The goal is to limit as much as opening the field of the choices of a designer and to help him in the development of his project by a certain degree of automation. We thus intend to create an intelligent ontology, equipped with reasoning structures. Ontology will include concepts on levels of definition based on an erudite knowledge of the buildings architecture (and of the territorial models in which they fit into). This ontology will understand concepts as rules allowing to implement reasoning during a process of project. Automating thus partially the process, once rules selected by a user, according to what he prefers, GED and TOP will transfer the rules on the project in progress. In the general principle, with each stage of the project, the designer can consult the ontology and find rules, that he can choose to apply, or not. For each selected option, a large amount of others are automatically eliminated, because they are incompatible.

Thus, by a set of successive transfers he orchestrates, the designer will evolve from a level of definition to another according to inferences suitable to order a process of innovating project (if not inventive), that's to say a process which doesn't reproduce what exists on the mode of the copy, but, on the contrary, assembles component elements, by partial copy, in order to produce new texts. It's not only a question of producing project 'in the manner of …', but also of offering a range of rules attached to levels of definition, allowing to accompany a process of project, by providing instruments of reflection, as possible reasoning, and as examples, starting then from components of specific languages, that will be composed to create new forms.

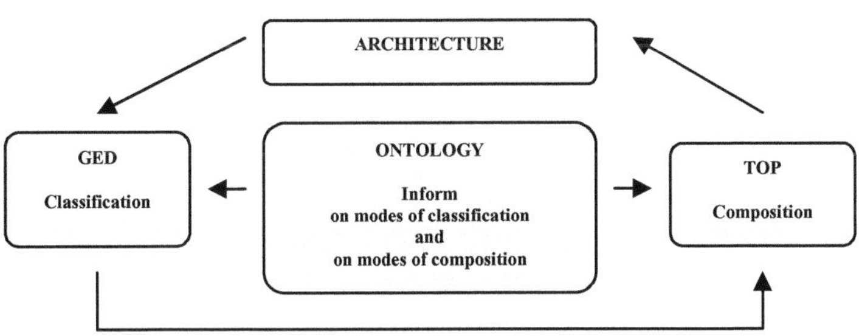

Fig. 1. From classification to composition

Any language by definition is a structure based on an economy of means; it's thus in itself a conceptualization of the world, a reduction of everything to its universal features. As it describes concepts of the common language, an ontology is a conceptualization of a field it describes; ontology releases from the features common to the language and the conceptualization of the thing. Any architecture is based on an order. We saw that when projecting, the architect passes from classification to composition; they engage two different forms of classifications, one, taxinomic, in hierarchy, and the other, in network. Because any classification proposes an order, an order that can follow various logics, consequently any classification is peculiar to an architect. The architect selects elements classified beforehand into a tree structure, which describes paradigms declined in paradigmatic elements; elements he assembles according to an order articulating a context in sentences, then in texts. Precisely because the context introduces distortions, the order of the composition finds its logic of setting in network of paradigmatic elements. A process of project is never linear; any designer operates multiple ways back.

Proposing a classification, an ontology is adequate to support a process of project; it's a receptacle in which one can compile, classify and make emerge various forms of order.

An intelligent ontology goes thus further than a thesaurus type one. It goes further, because it opposes subjectivity to objectivity, in the sense that it doesn't claim to describe the world as it is, in abstracto of the subject that conceives it; it opposes to an order based on classification, an order based on reasoning, because it wants to be the support of forms always innovating if not inventive; it thus opposes inferential mechanisms (survey by rules; logic of proposals) with non-inferential mechanisms (logic of classes). But it also includes a consultative function, it thus includes referential mechanisms. It compiles a series of references, which inform on the various levels of definition of a concept. We thus wish that the ontology[3] offers a support of reasoning completely similar to those used by a designer in a natural way.

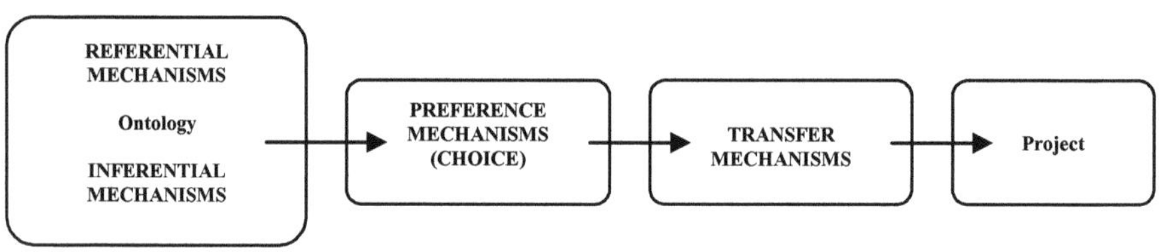

Fig. 2. From ontology to project

[3] To elaborate the ontology, we'll lean on *Protégé 2000*, software that permits to build ontology and where it's possible to program new functionalities.

4. From a bifacial to a trifacial concept, from Saussure to Peirce

Because the knowledge of the being is based on subjective interpretation, an erudite ontology rests on hermeneutics conception. According to Eco, *'as soon as it makes irruption in front of us the being causes interpretations'* [04]. The hermeneutics principle wants that there are no facts, but only interpretations of facts. Because everything implies a choice, a point of view independent of what are things, the being can be only what is said in multiple ways. When projecting, the architect interprets a program and a place; when we read his project, we interpret in our turn what he says to us. The cycle of interpretations is infinite. An ontology is thus only a possible interpretation of a field of the being in the world that it conceptualizes.

The interpretation of architectural texts, of their language, starting from the analysis of plans, or from the discourses produced on the building, as metalanguage, allows us to build our models. The knowledge which results from this is a complex and hypothetical construction, in the sense that it doesn't offer only one point of view on an architectural production. It puts at the disposal of a user a semiotic and semantic network in which he moves, guided by his own subjectivity (fixed by his lived, his personal aspirations), with which he builds his own hypothesis, and his own conception of the relation that the human being can maintain with the world.

This approach of the problem supposes to go beyond the definition of Saussure, from now on unsuited to reach our scientific objectives [02]. Indeed, for Saussure, bifacial, the sign is the product of the relation of a signified (*concept*) and a signifier (*acoustic image,* or iconic representation allowing to constitute the sign), relation whose emerges the meaning of the sign (I stress that Saussure uses a definite article). His definition takes into account only the object and not the subject; however it's the latter that interests us, the taking into account of the subject implies a plurality of possible signification for the same sign. We will thus starting from a definition of the conceptual sign opened to the insertion of new definitions (each definition bringing a new meaning to a concept), where consequently the signified as well as the signifier can be enriched by new signification.

The definition of the sign we'll lean on from now on borrows from Peirce [05]. The trifaciality of the sign adds thus a third dimension to the definition of the sign of Saussure, that of the interpreter of the sign. Postulating that between the object and the sign intervenes the logic of the subject, according to Peirce, the understanding of a sign is never the same for everyone of us. The understanding of the object cannot be done without the taking into account of the subject, and the signification which results from it is not passively given; it results from an inventive mediation of the interpreter, of his conscious will. However, it's because architects can extend the definition of a concept that invention can occur. It's really because any interpretation is primarily a hypothesis according to a rule applied to a case, and consequently inferential ('if x... then y...'), that the representation of an object can be transformed.

5. Methodology

The research we undertook here doesn't aim neither at exhaustiveness, nor at the description of the greatest number of concepts, but rather to study, through a restricted number of cases, how can be solved the general problem. A solution valid for some cases is enough; ontology could be updated thereafter by successive incrementations. To achieve these goals, we started to work in three directions.

First direction of research – calculation on concepts

Our work found a field of application within the framework of WINDS [06], a vast European research program relating to the creation of a virtual university in the field of architecture. Each participant had to formulate theoretical courses and adequate practical exercises with a teaching via Internet. It was imposed to index the keywords structuring the courses and exercises, as their relations, according to five kinds of links. The classification of the concepts in a 'hyper-structure' aimed to give access to various lessons according to particular fields of interest. As teaching provides the base of the vocabulary and of the reasoning students will apply during their career, WINDS offers an adequate framework to find the concepts of our ontology. It's a question of selecting some concepts and studying their multiple significations in the context of various lessons.

These concepts and their relations can be translated into graphs. Our first direction of research thus consists in making calculations of distance on the relations between these concepts, using GED, and interpreting the results. We can know particular properties of the relations between the WINDS concepts and turn them to account during the construction of the ontology.

Second direction of research – put in parallel of a process of project and an ontology

The second direction seeks to put in parallel a process of project and an ontology, so that the ontology can feed a project in progress. From the point of view of the content, our ontology is conceived not only as a data medium, but also as a help with the reasoning; it will be moreover equipped with normative rules related to a reasoning, rules which are those of a model.

When evolving toward a solution, the thought of the architectcharacterizes by ceaseless going on and going back. With each stage of the process, the designer will connect on the ontology and seek information and assistance about the reasoning, according to a way defined by himself.

Third direction of research – analysis of texts

While setting out again of the data of a semiotic articulation, we intend to highlight various levels of definition in the production of an architectural concept. It is necessary to distinguish, like does Bonfantini [05], the mechanisms from which rise innovation and invention. If each project innovates, in the sense that it is never the certified copy of another one, it is not therefore always carrying an invention. Innovation emerges from the recombining of existing bits; it is always only partial.

It is different with the invention. Masters are architects who, at a certain point, are able to transgress the codes in force to propose something else, a new way of life, a new ethics, a new esthetics. As stresses Saussure [02], language is alive, it follows a diachronic evolution. The invention in architecture concerns the extension of a concept, with one, or more, of its definition levels, if not the creation of a new concept. This mechanism follows rules similar to those of the natural language, where, unceasingly, new words come to supplement the existing vocabulary, accompanying thus progress in various fields of the social knowledge.

Then, succeeding to an analysis on the language, the third direction of research consists of an analysis on the metalanguage which frames the buildings production. The analysis relates to the speeches produced about their building by the Masters themselves and also about their general theories; as well as the building, the text is a memory of the intellectual advance of the construction, of the invention of a language. The emergence of a new concept or its extension is the result of a reasoning which guided the architect towards a solution. An analysis on the metalanguage permits to seize how a reasoning 'is constructed', reasoning from which the rule emerges. Any rule is by definition inferential.

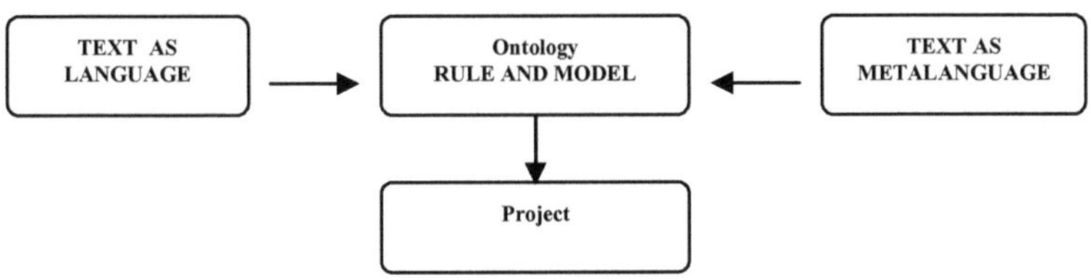

Fig. 3. Ontology between language and metalanguage

6. The seven levels of definition of an architectural concept

For each architectural concept, there are various definition levels, all starting from a definition of the language. The first level proceeds by reduction of information; it describes universal features of a concept. This level is closed; the definition cannot be supplemented, because, in this case, it would open another concept.

The second level treats of the form of the content, of the concepts and of the way they can be classified. It connects concepts according to reasoning's suitable for a natural logic [07]. This definition level is opened to extensions; new functions or properties can supplement the universal definition.

The third level relates to the shape of the object. A paradigm describes possible geometries that can incorporate an object. In the same way as in the natural language, where paradigmatic elements can replace the ones the others inside a sentence, paradigmatic elements describing forms can take place inside an architectural syntagmatic chain. Paradigm and syntagm thus allow an economy of the language.

The fourth level relates to the substance of the object and works according to the same principle as the preceding level of definition. But the form already contains information on the substance; certain forms are adequate only for certain materials and thus limit the field of applicable materials.

The fifth level relates to the dimensions an object can take. Variable according to architects, metric contributes to define the style of an architecture.

The sixth level relates to references. In architecture, the reference is one of the mechanisms producer of meaning, as one of the mechanisms decoder of meaning. References will be attached to concepts in textual or iconographical form. To connote its building, to add it more meaning, the user can seek in a particular semantic field he will report next onto his building.

The seventh and last level contains normative rules allowing to implement a reasoning, starting from the three known inferential forms. Any inference has necessarily three terms, an antecedent (a case), a consequent (a result), an implication (a rule). According to their order, one deals with three different forms of argumentative logic, either deduction, or induction, or abduction. According to Peirce [05], abduction is the only form that allows invention.

Rules are based on the one hand on theories formulated by recognized authors; they are moreover based on our works of modelization of architectural languages, thus allowing the implementation of a project in a certain style. Rules can be applied as much to phases of classification as of composition.

If the system we develop allows to implement reasoning and to automate partially the process of design, it doesn't remain about it less than in last spring, it is the designer who will determine what will be his project; according to his own conception of the model, he will be able to build his own language.

7. Concepts, metaconcepts and connotation

We choose Protégé as software to build our ontology; Protégé allows to classify the concepts in a tree structure based on inclusion and exclusion relations. Moreover, according to a logic of classification in network, Protégé allows to connect concepts belonging to any branch of the tree via properties, and restrictions on these properties, thanks to which it is possible to describe or define the relevant features of the numerous facets of a concept.

The interpretation of Vitruvius' definition of architecture, by Pierre Pellegrino [08], has given a starting point to the organization of the concepts of our ontology. Here is this definition: *'In architecture, as in any other* ***science****, two things are noticed: that which* ***is meant****, and that which* ***means****. The meant thing is the stated thing about which one speaks, and that which means is the demonstration that one gives through the reasoning, supported by science'.* According to Pellegrino, for Vitruvius, theory is based on a double system, a metasemiotic one and a connotative one. The metasemiotic level treats of a metalanguage, which words are metaconcepts contributing to describe the building (*'the stated thing about which one speaks'*) as an instrument with an utility, where each element is selected, connected with others, shaped and sized considering its utility. Metalanguage is therefore a language allowing to speak about an other language, in this case the code according which is recognized the utility of an instrument.

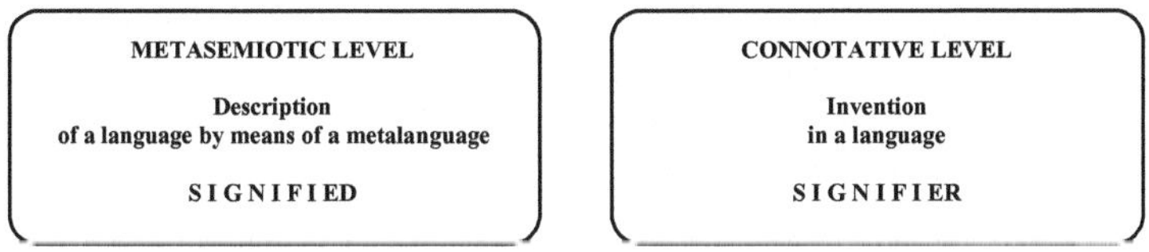

Fig. 4. Metasemiotic level and connotative level

Each architect can also use metaconcepts to add more meaning to elements of his project. One passes then at a connotative level. One leaves the universe of description, of 'speaking of architecture' for the one of representation, of 'speaking architecture'. It is then necessary to understand how works the mechanism of connotation in architecture. At a connotative level the designer uses universal metaconcepts, but of his own. For example, each architect can use the metaconcept of proportion; but it is according to a specific way to consider the question of what is 'good proportion' that each one will connote his project. Other example, an architect, trying to open interior space on exterior space, who actualizes only some layouts of the geometrical figures of his plan (here squares); complete figures, in their virtuality, connote therefore the opening of the space and its relative closing. One can notice that complete figures are connoting differently the openings depending on whether sides or angles are erased. So, a specific manner to use metaconcepts, in this example the game between the couple actual/virtual, as elements of the project, can be a connotation of the reasoning made in the project.

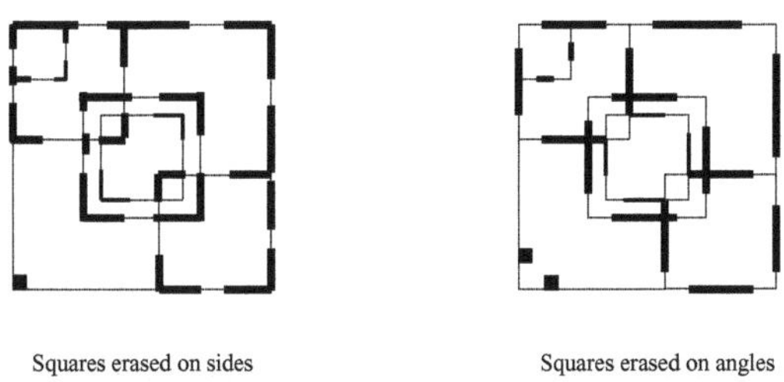

Fig. 5. Example of connotation of the reasoning

This being, the process doesn't end here. Some metaconcepts, in specific languages, classified as connotation, become in turn universal metaconcepts; they can be used later to connote different architectural productions, in other languages. We are so inside a circular mechanism. In the ontology, one must thus classify concepts in the same time as invention in a language describing the object of the project, as metaconcept allowing to speak about this language as an object of the project, and finally as connotation of this same language.

It is being connoted by use of metaconcepts that concepts transform, evolve, decline in specific features. Both levels are complementary, take on and back meaning one relating with the other.

Thus, the analysis of Vitruvius' definition allowed us to define three very general branches, describing, for the first, objects of the project, or instances, for the second, all the metaconcepts of the metalanguage, and for the third, a series of metaconcepts used as particular features in languages.

8. Examples of inferences

Here is what should be necessary to bring in Protégé to build an ontology suitable for an epistemology of architecture. The stake consists in passing from a descriptive ontology, conceived as an intelligent library describing classes of reference to make project in the manner of... such-and-such architect, to an ontology that gives an account of reasonings. In Protégé, concept A is linked to concept B via a propertie. It's what one can call an ontology with a bi-dimensional structure; propositions add up and link up by successive transfers. That allows to describe components of languages, to illustrate them with texts or pictures; it corresponds to one of our objective: a consultative ontology, offering references. To reach the second objective, we have to build an ontology giving an account of inferences, that's to say of rules in a language. That supposes inferences calculated by an interpreter and endowed with three principal forms: deduction, induction and abduction, each one integrating the three sequences 'antecedent, case and consequent', suitable for invention and architectural reasoning. The ontology could thus give an account of sequences of propositions, from which follows a result, what one can define as a third dimension of the ontology[4].

Here is, to conclude this article, an example of specific inference we wish to implement on our ontology, in the way to define a concept as product of the reasoning of an interpreter. The concept of 'plan libre' doesn't mean the same thing depending on whether it's produced by L.I. Kahn's or Le Corbusier's mind. Their respective reasoning, when they are trying to convince us about the indisputable quality of the spaces they are projecting, is related here in form of theorem.

The concept of 'plan libre' is a Le Corbusier's invention [09]; he sought convincing arguments to persuade the public of the superiority of the modern house's plan (*'plan libre'*) in comparison with the traditional stone-built house's plan (*'plan paralysé'*), as it was usual. According to him, in the traditional house, load-bearing wall technique paralyzes the plan; for reasons due to statics and resistance of materials, the plan is constrained to repeat identically at each floor, although in reality, ideally, for different activities should correspond rooms with various shapes and sizes. In the modern house's plan, Le Corbusier resolves the question moving the load-bearing function of the wall onto the column. He can thus substitute them by dividing walls (that aren't supporting anything), that can freely inhabit the space of the plan.

---

[4] We are actually reviewing the different builts-in already available on Protégé.

According to Le Corbusier

If SolidColumn is Load-bearingSpace
And if Load-bearingSpace isConditionOf PlanLibre
*Then*SolidColumn isConditionOf PlanLibre

Starting again from Le Corbusier's theory, next leaving explicitly the Master, Kahn try to convince us that it is at the end of a very different reasoning, that it is possible to free the space of the plan [10]. Through a metonymy, Kahn relates traditional house's plan (considered as a whole) on the column (considered as a part), column he envisages as a room, that's to say endowed with walls, empty and usable. Kahn puts the accent on the column as serving space and not anymore as load-bearing space. It's by means of the fusion of physical and functional dimensions that plan really becomes frcc.

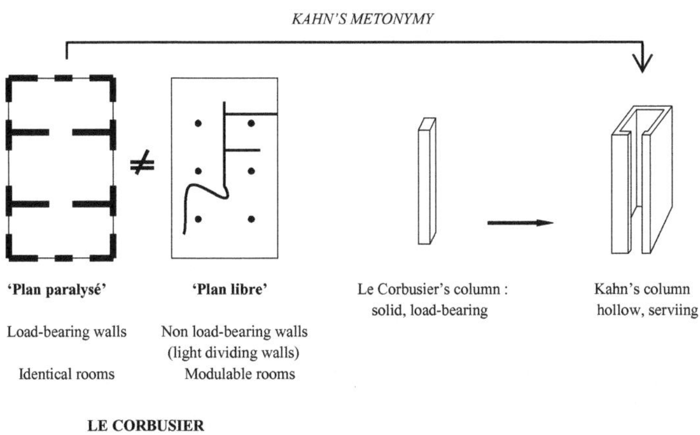

Fig. 6. Plan libre according to Kahn and Le Corbusier

According to Kahn

If HollowColumn is ServingSpace
And if ServiingSpace isConditionOf PlanLibre
Then HollowColumn isConditionOf PlanLibre

According to the metaconcepts attached to 'column', in this case if it is a load-bearing or a serving space, that reports on the properties of the column (solid or hollow), implications about what is 'plan libre' are totally different and found very specific dimensions of the structure of the langage of both architects.

9. Outlines

The explicit competence of an architect is to create forms. Its implicit competence is to analyze the process he implements, to seize how he connects contained forms with container forms, how he links at the same time as he excludes, how he seizes dimensions of a context to create forms, to give a support to our existences. The transition from the natural intelligence to the artificial one can be done only if based on semiotics models which precisely can account for the intellectual advance from a desire to a built object. For this reason, an ontology conceived like an operative structure, based on inferential mechanisms, and consultative, based on referential mechanisms, allows us to approach more and more our objectives.

References

1. Coray, D., Pellegrino, P., Jeanneret, E. P., and alii: La conceptualisation de l'espace en architecture I and II, CRAAL-FNSRS, Berne (1995 et 1998)
2. Saussure, F.: Cours de Linguistique générale [1915], Payot, Paris (1972)
3. Pellegrino, P., Jeanneret, E. P.: WINDS CONCEPTS, Interdisciplinary research within the framework of an European program (COST C-21), TOWNTOLOGIES (2006)
4. Eco, U.: Kant et L'ornithorynque [1997], fr. transl. Grasset, Paris (1996)
5. Bonfantini, M.: La semiosi e l'abduzione, Studi Bompiani, Milano (1987)
6. Pellegrino, P., Jeanneret, E. P., and alii: WINDS, Web Intelligent Tutorial Design System, Ancona (2000)
7. Grize, J.-B.: Logique naturelle et communications, P.U.F., Paris (1996)
8. Pellegrino, P.: 'The semiotics of architecture : A Heritage' [1998]. In :Das Europaïsche Erbe der Semiotik; The European Heritage of Semiotics, Dresdner Studien zur Semiotik Bd.2, Dresden (2004)
9. Le Corbusier: Précisions sur un état présent de l'architecture et de l'urbanisme, Fréal, Paris (1960)
10. Rivalta, L.: Louis I. Kahn, La construction poétique de l'espace, Le Moniteur, Paris (2003)

Author Index

MIX
Papier aus verantwortungsvollen Quellen
Paper from responsible sources
FSC® C105338

If you have any concerns about our products,
you can contact us on
ProductSafety@springernature.com

In case Publisher is established outside the EU,
the EU authorized representative is:
Springer Nature Customer Service Center GmbH
Europaplatz 3, 69115 Heidelberg, Germany

Printed by Libri Plureos GmbH
in Hamburg, Germany